MATH/STAT
LIBRARY

Douglas Rayner Hartree

His Life in
Science and
Computing

Douglas Rayner Hartree

His Life in Science and Computing

Froese Fischer
Vanderbilt University, USA

World Scientific

NEW JERSEY • LONDON • SINGAPORE • SHANGHAI • HONG KONG • TAIPEI • BANGALORE

Published by
World Scientific Publishing Co. Pte. Ltd.
5 Toh Tuck Link, Singapore 596224
USA office: Suite 202, 1060 Main Street, River Edge, NJ 07661
UK office: 57 Shelton Street, Covent Garden, London WC2H 9HE

British Library Cataloguing-in-Publication Data
A catalogue record for this book is available from the British Library.

DOUGLAS RAYNER HARTREE: HIS LIFE IN SCIENCE AND COMPUTING

Copyright © 2003 by World Scientific Publishing Co. Pte. Ltd.

All rights reserved. This book, or parts thereof, may not be reproduced in any form or by any means, electronic or mechanical, including photocopying, recording or any information storage and retrieval system now known or to be invented, without written permission from the Publisher.

For photocopying of material in this volume, please pay a copying fee through the Copyright Clearance Center, Inc., 222 Rosewood Drive, Danvers, MA 01923, USA. In this case permission to photocopy is not required from the publisher.

ISBN 981-238-577-0

Printed in Singapore by World Scientific Printers (S) Pte Ltd

Preface

Anyone familiar with quantum mechanics will have heard of Douglas R Hartree and his self-consistent field method. Though Fock pre-empted Hartree's work on equations with exchange, Hartree later extended the equations to include configuration interaction. In doing so, he played an important role in laying the foundation not only of modern day atomic physics but also of quantum chemistry. Not nearly as well known are his contributions in other areas. In fact, it has been said that Hartree only had this one idea, churning out atom after atom. This was hardly the case. Early in his career he became interested in radio wave propagation and derived the equation for the index of refraction that in atmospheric physics today is called the "Appleton-Hartree equation." During World War II, he applied the self-consistent field idea to the study of the cavity magnetron then part of radar development. Computationally, this was a far more challenging problem than atoms because the geometry of cavities and slots precluded the assumption of radial symmetry, requiring the solution of two-dimensional partial differential equations. From these computations, he made a number of discoveries, one being the "Hartree condition" for the operation of a magnetron. There were others.

The common denominator of Hartree's research was differential equations, either ordinary or partial. He combined his wide range of interests with a talent for problem solving. Given a problem, he would express the phenomena investigated in the form of one or more differential equations, devise a scheme for solving the equations in their simplest possible form, and perform the calculations. In the 1930's this led him to the differential analyzer and its application to a variety of problems, the most important being in control theory and fluid dynamics. Hartree knew the accuracy of the differential analyzer was insufficient for most atomic structure work, but he believed there were many industrial applications where great accuracy was not needed and the differential analyzer would be a very welcome tool. In this he was correct. During the war, he dealt with problems that were partial differential equations, and in later years, Hartree was greatly interested in equations arising in hydrodynamics. Because he believed in solving a problem with any tool available – slide-rule, desk calculator, or differential analyzer – he acquired great computational skills and knowledge of what was later called numerical analysis.

Hartree did not hide his "love of arithmetic." This was not universally greeted with admiration or acclaim. Slater, in his autobiography, classified scientists as the "hand-waving magical type" or the "matter-of-fact type." He placed Hartree in the latter category, saying that the hand-waving-magical type considered him a "mere computer," yet he made a greater contribution than most of them.

So why would someone under those circumstances take such an unconventional stand and pursue "arithmetic?" I believe the answer lies in the unusual family background, and the Bedales school Hartree attended prior to University. Both nurtured the development of his unique talents and at Bedales he learned to consider his role in society. For this reason, the present biography starts with the Hartree family.

Because the extent of Hartree's contributions is not well known, the rest of the book attempts to chronicle his science as well as his life. Many of his publications are not readily accessible today, so a summary is appropriate. Hartree did not consider himself important and did not keep correspondence, neither with scientists nor with family members. Fortunately, some archive collections have kept Hartree's letters – Edward V Appleton, Niels Bohr, Vannevar Bush, Robert B Lindsay, Fritz London, Rudolf Peierls, John C Slater, Ivar Waller – but particularly before World War II, letters were written by hand without carbon copies so the "conversations" in many letters are somewhat one-sided. As a document of his activities, it has been necessary to rely on Hartree's own publications supplemented by letters, as available, and documents from the Public Record Office, London, relating to war time activities.

Hartree spent the 1920's in Cambridge, first as an undergraduate, then as a graduate student and research Fellow at the Cavendish laboratory. During this time, Ernest Rutherford was Professor of Physics and Ralph H Fowler, a lecturer in Mathematics, supervised those more theoretically inclined. Rutherford attracted many physics students from around the world to the Cavendish. The students of these two men were the "community" with which Hartree identified. The science world at that time was much smaller than today so the same "cast of characters" will appear in different chapters of this book, ranging from World War I, quantum theory, World War II, and the emerging computer era.

In placing Hartree's early work in context, it has been necessary to outline some of the developments in quantum theory, but I do not consider this to be a history of quantum mechanics. John Slater, who was more intimately involved in some of the theoretical ideas, has written an autobiography, *Solid State and Molecular Theory: A scientific biography*, which, in spite of its title, presents an account of historical developments, and could be an excellent companion for readers desiring a more detailed account still at a simple level.

In discussing Hartree's science, I made the decision that, when necessary, differential equations would be included, but often they only show the nature of the problem in a concise way. The detailed meaning is of lesser importance. His papers are not cited here, but a complete list of his publications is included at the end of the book.

Finally, let me say that, as Hartree's student, I unfortunately did not get to know him well. I never attended any lectures that he presented. Shortly after my arrival at Cambridge, Hartree was getting ready for his year at Princeton. When he got back, he understandably was exceedingly busy and soon thereafter, I was getting ready to return to Canada. The subsequent year we corresponded and Hartree arranged for my dissertation defense at the University of Toronto. Six months later he died unexpectedly. Thus not much of this book is from personal recollections.

This biography could have been a lot different had it been started sooner. I did contact Bertha Swirles Jeffreys and Jack Howlett in 1999, shortly before they died, getting some valuable information. I had one letter from David Myers who had been Dean of Engineering at the University of British Columbia during my time there, but we were just establishing contact when he too passed away. I was not aware how intimately he had been associated with Hartree and his work on the differential analyzer until starting this biography.

However, I was fortunate in other respects. I got to meet Hartree's first PhD student, Arthur Porter, who alerted me to the importance of their work on control theory. In the last few years, a great deal of information has become available on the Internet that was a great asset. The "Nobel e-Museum" has a wonderful selection of biographies and lectures (http://www.nobel.se). Anyone not familiar with computing devices can visit the museum of HP calculators (http://www.hpmuseum.org) and get more information about slide-rules, calculating machines, and calculators than they might wish to know. And there are many others. In some instances, the web sites were a source of information, such as http://www.alanturing.net which has on-line copies of the minutes of NPL meetings in which Hartree participated, but in others, such as the one for calculators, they illustrated the fact that readers can readily find additional information of a general nature. Unfortunately, web sites come and go, therefore they have not always been included here as references.

The Internet also made it possible for me to contact the Hartree family. I was desperately searching for pictures of Hartree that are few in number. I knew Hartree had collaborated with Robert Bruce Lindsay at Brown University. While searching the Brown University site, I came across the address of his son-in-law who provided me with the address of Hartree's daughter, Margaret Hartree Booth. As soon as we made contact, the Hartree family (Margaret, Oliver, and Richard) graciously shared with me the family pictures included in this book.

Several factors motivated me in undertaking this project. I noticed that the Emile Segrè Visual Archives had 36 (it now has 53) pictures of Dirac and only one of Hartree, and that is a group photo with Einstein. This indicated to me that pictures of Hartree were badly needed for an accurate historical "picture gallery" of the scientists of his time. Then in searching the McGraw-Hill online encyclopedia, I found Hartree's name associated not only with atomic physics, quantum chemistry, band-theory of solids, the nuclear molecule, and nucleon-nucleus scattering, all of which I was aware of to various degrees, but there also was the Hartree equation in electronics! This was totally new to me. I later found it to be related to the operation of a magnetron. At the same time, there was no mention of his work with the differential analyzer though the biography of Vannevar Bush mentioned that a large differential analyzer had been completed in 1935 at Manchester University and was now an exhibit in the Science Museum, London, without mentioning that the person to do so was Hartree! All this suggested to me that a biography was needed to describe the many facets of his work.

In this biography, I have attempted to gather available information about Hartree and his work as well as give a sense of the times in which he lived.

Charlotte Froese Fischer

Acknowledgments

This biography has relied on many valuable sources of information.

The Hartree family – Margaret, Oliver, and Richard – provided me with the pictures included in this biography, unless indicated otherwise. Christ's College, Cambridge, was most helpful in getting me started, sending me Margaret Hartree Booth's memoirs, and obtaining letters from Hartree to Edward V Appleton from the Special Collections, Edinburgh University Library. Dr Michel Godefroid of the Free University of Brussels, shared with me his collection of articles on the Meccano differential analyzer. Dr Jon Agar, of the Centre for the History of Science, Technology and Medicine, University of Manchester, provided me with some difficult to find Hartree publications and an exceptional picture of Hartree demonstrating his differential analyzer. Dr James Peters, Manchester University Archivist, clarified Hartree's involvement in the development of the Faculty of Music.

The Niels Bohr Library, Center for History of Physics, American Institute of Physics, College Park, MD through their oral history interviews, provided notes of a conversation with Elaine Hartree, transcripts of interviews with Sir Nevill Mott and John C Slater as well as the Robert Bruce Lindsay autobiography. Through the Archives for the History of Quantum Physics program, I received letters from Hartree to Samuel Goudsmit, Robert B Lindsay, and Niels Bohr. The Niels Bohr Archive in Copenhagen provided letters to Niels Bohr, Oskar Klein, and Léon Rosenfeld. The Center for History of Science, of the Royal Swedish Academy of Science, Stockholm, kindly furnished copies of correspondence between Hartree and Ivar Waller. Letters between Hartree and Fritz London are part of the Fritz W London Papers in the Duke University Archives whereas the correspondence with John C Slater is part of a collection found in the library of the American Philosophical Society, Philadelphia. I am grateful to all these archives for the information they have provided. I am also grateful to Peter Thorne, Leader of CSIRAC (Australia's first automatic computer) history project for searching the CSIRO archives for memos relating to Hartree's visit to Australia in 1951.

In discussing research in a variety of areas I have relied on expert advice from others – Arthur Porter for research on control theory, Ronald Lomax for the fundamentals of the magnetron, Sir Maurice Wilkes for computer developments and Appleton's work on radio wave propagation, and Harry Huskey for his participation in the ACE project. I also wish to acknowledge my appreciation of those who submitted a brief statement on how they happened to become Hartree's students – Arthur Porter, John Crank, Nicholas Eyres, Sir Aaron Klug, Roy Garstang, Donald Leigh, David Mayers, and Ronald Lomax.

Special acknowledgments are due to Roy Henry Garstang, also one of Hartree's students, now at the University of Colorado. He started by providing me with copies of most of Hartree's published papers. He also sent me immense amounts of background information, biographies, and summaries that were an excellent starting point for writing this biography. He was my expert consultant on Cambridge tradition. Having obtained first an undergraduate and then a graduate degree at Cambridge, he had a much better appreciation of the University and its faculty than I did.

Last, and by no means least, I wish to acknowledge the many contributions made by Richard Hartree. He provided me with invaluable assistance by pointing me in appropriate directions for expanding my sources of information, visiting the Public Records Office in London for reports that could only be obtained in person, and for answering many queries about the Hartree family and his father. He also was a strong advocate for "presenting the story." For someone accustomed to technical writing, this was the most difficult challenge of all.

Douglas Rayner Hartree (1897 - 1958)

Picture of Douglas Hartree, taken about 1939.

Chronology of events

1897 Born in Cambridge, England, on March 27
1910 Attended Bedales School, Petersfield
1915 Began University studies, St John's College, Cambridge
1916 Joined the A V Hill team (anti-aircraft section, Ministry of Munitions)
1919 Returned to University, St John's College, Cambridge
1922 Awarded BA degree
1923 Married Elaine Charlton
     Elected Fellow of the Cambridge Philosophical Society
1924 Elected Fellow, St John's College, Cambridge
1926 Awarded PhD degree
1928 Elected Fellow, Christ's College, Cambridge
     Visitor, Niels Bohr Institute for Theoretical Physics, Copenhagen
1929 Beyer Professor, Chair of Applied Mathematics
     University of Manchester
1932 Elected Fellow of the Royal Society, May 5
1937 Chair of Theoretical Physics, University of Manchester
     (on leave 1940-45, Ministry of Supply)
1939 President, Manchester Literary and Philosophical Society
1945 Professor of Engineering Physics, University of Manchester
1946 Plummer Professor of Mathematical Physics, University of Cambridge
1948 Acting Director for several months, Numerical Analysis Institute, National Bureau of Standards, UCLA
1958 Died February 12

Hartree family tree*

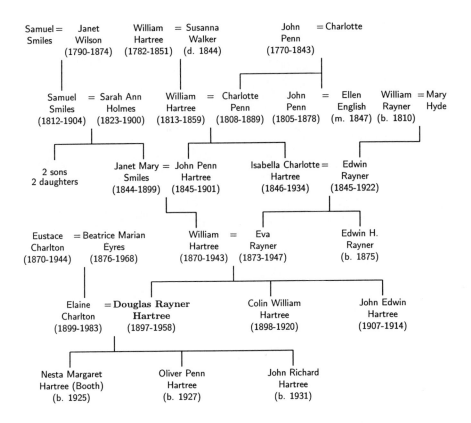

*Not all the spouses, children, and grandchildren are shown here, only those relevant to this biography. A more extended family tree of the Smiles family can be found in *Samuel Smiles and his surroundings* 1956 (London: R Hale), written by Smiles' granddaughter, Aileen Smiles.

The author is grateful for the assistance received from the Hartree family, namely Nesta Margaret Hartree Booth, Oliver Penn Hartree, and John Richard Hartree, in preparing this family tree.

Hartree family tree

Contents

Preface v

Acknowledgments ix

Douglas Rayner Hartree xi

Chronology of events xiii

Hartree family tree xv

1. The Hartree family 1

2. Education, World War I, and marriage 9

3. Early research at Cambridge University 23

4. The new quantum mechanics 33

5. Advances in atomic theory 47

6. Radio waves in the atmosphere 65

7. Professor at the University of Manchester 73

8. The differential analyzer 85

9. Control theory and industrial applications 103

10. Laminar boundary layer theory 109

11.	Arrangements for war	115
12.	Wartime service	123
	12.1 Servo Panel	123
	12.2 The differential analyzer job shop	124
	12.3 Magnetron research group	136
13.	Dawn of the computer era	145
14.	Returning to Cambridge	161
15.	Summers in North America	171
16.	Mathematical Laboratory, numerical analysis, and teaching	177
17.	A trip to Australia	181
18.	Atomic structure research using EDSAC	187
19.	The final years	195
20.	His legacy	201
	20.1 Bertha Swirles Jeffreys	201
	20.2 Arthur Porter	201
	20.3 Jack Howlett	204
	20.4 John Crank	204
	20.5 Nicholas Eyres	205
	20.6 Sir Aaron Klug	205
	20.7 Roy H Garstang	207
	20.8 Donald C F Leigh	207
	20.9 David F Mayers	208
	20.10 Ronald J Lomax	209
	20.11 A personal tribute from the author	210
Hartree's publications		213
Index		221

Chapter 1

The Hartree family

Douglas Rayner Hartree was born in Cambridge, England on March 27, 1897, to William and Eva Hartree. He was their first-born and was given his mother's maiden name as middle name. They were a middle class family, comfortably off with a largish house and a complement of domestics.

William Hartree was named after his grandfather and great grandfather, all with the name "William Hartree" (see the family tree on page xv). Grandfather William Hartree [1] was born in 1813 at Rotherhithe, in Southeast London (on the south side of the Thames, upstream from Deptford and Greenwich) and educated at the Merchant Taylor's School in nearby Lewisham. Demonstrating an early talent for the mechanical, he became a pupil of John Penn & Sons in Greenwich [2], a partnership primarily of John Penn and his son, both also with the same name.

John Penn Sr (1770-1843) moved to London in about 1793 and started a business as millwright at Greenwich where he soon acquired a reputation for the construction of flour-mills. Among his many improvements was the introduction of cast iron instead of wood for the framing. He was known as an "Ironmaster" because the works operated a foundry. In 1825 he began to turn his attention to marine engines, the first of which was for a steamer running from London to Norwich. His son, John Penn Jr (1805 - 1878) entered his father's works at an early age where he quickly developed into a mechanical genius. He soon took a leading part in his father's business, so that it became difficult to know what share of the many contributions to the firm of John Penn & Sons came from the father and what share from the son. By the time of the father's death, the son had practically taken over the management of the firm.

At the same time, William Hartree worked his way up in the firm to become works manager and eventually partner, taking a very active and prominent role in the management of the firm. He also became a friend of the Penn family and in time married Charlotte Penn, John Penn Jr's sister.

The Penn's marine engineering works was an active business on the Thames when London was the largest port in the world. After the death of the father, the son continued introducing innovations related to screw propulsion. The engine works was established on Blackheath Road, Greenwich, and marine boilers were produced at the firm's works at Payne's Wharf, Deptford. Between 1845 and 1878, eight UK patents were taken out in the name of John Penn Jr with his two partners (one being William Hartree) included in the most important ones. By the time of his death in 1878 the firm had engined some 735 vessels. The engines and boilers for HMS Warrior, a ship that transformed concepts of naval warfare when she joined the fleet in 1861, were built by Penn's engineering works. The powerful steam screw propulsion meant that she could outrun any ship afloat. This was the first British iron warship and is now preserved in Portsmouth. As ships grew larger, the Thames was less suitable for shipbuilding and the company declined being finally wound up and sold in 1901.

Because of their success, the Penn family was one of substantial standing in the community and local society. John Penn Jr and his wife were most hospitable by nature and for many years entertained a large number of guests at their house on Belmont Hill, Blackheath, known as "The Cedars." Upon the death of John Penn Jr, his widow erected the Penn Almshouses in his memory. They can still be seen in Greenwich today, though somewhat altered after the World War II damage. The name of John Penn is commemorated by John Penn Street, also in Greenwich.

William Hartree and Charlotte Penn shared in the good social and financial position of the Penn family. They had five children, of which only John Penn Hartree (1844-1901) and Isabella Charlotte Hartree (1846-1934)) are included in the family tree. In 1851 William Hartree became an Associate of the Institute of Civil Engineers and in 1857 a Member, in recognition of his "pupilage in the works of Messrs John Penn & Sons, the direction of these works, and his year as Partner." His good fortune was short lived. On a work related trip in 1859 William contracted a chill and died of pneumonia a few days later at the relatively early age of 45. His wife outlived him for thirty more years during which time she continued to associate closely with the Penn family. William Hartree had been much devoted to literary pursuits and was a connoisseur of rare and contemporary books. He was a collector and during his lifetime assembled a valuable and extensive library. Many books were from limited printings, others were old (some from as early as 1476) and very rare. Among his collection was a most extensive and important collection of engravings from the work of Sir Peter Paul Rubens, a collection of some 1,350 engravings arranged and mounted in five large volumes. At the time of his wife's death, the library was "sold by auction by Sotheby, Wilkinson & Hodge on Thursday, the 10th day of July, 1890 and Seven following days" [3]. The catalogue itself contains 2,543 entries. William Hartree must have been a most remarkable engineer.

The Hartree's lived on Morden Hill off Lewisham Road, about a mile from the home of Samuel Smiles, a social reformer and a prolific writer of biographies, copies of which were part of the William Hartree library.

Samuel Smiles [4; 5] was born on December 23, 1812 in Haddington, Scotland, a small town situated about 16 miles east of Edinburgh. He was the third child of eleven surviving children of Samuel and Janet Smiles. After being educated at Haddington grammar school, Samuel was apprenticed to a firm of medical practitioners. Already at that early age, he was a strong believer in "self-improvement" (education) and, through a great deal of determination, obtained a medical diploma in 1832 from the University of Edinburgh. He returned to Haddington, settling in as a general practitioner. But the competition for paying customers was great. To supplement his income and apply his spare time to self-improvement, he started writing and giving popular lectures on chemistry, physiology, and the conditions of health. His first book, *Physical Education or the Nurture of Children*, was well received. The 750 copies that he had published at his own expense (1836), were soon sold: a second edition was printed in 1868. In 1838 he left Haddington for a foreign tour to broaden his knowledge.

Coming back to England, he obtained a position as editor of the *Leeds Times*. Here Smiles combined his editorial duties with political activities in support of liberal causes. He urged the extension of suffrage and the "leveling up" of the working class by education and improvement. He also interested himself in industrial organization and the progress of mechanical science. In 1842 he resigned the editorship of the *Leeds Times*, devoting himself to popular lecturing and writing. He also became interested in railways and served as Secretary for a number of railway companies, first in Leeds (1845-54) and later in London (1854-66). It was then that he moved to Blackheath, a suburb of London.

In 1845, Smiles delivered a lecture to a small society at Leeds which was well received. He enlarged on the lecture, out of which grew the book *Self-help, with Illustrations of Character and Conduct* [6]. The book contained examples from all walks of life, on subjects ranging from art and music to industrial engineering, of individuals achieving success by overcoming great adversity. But Smiles also had a way of presenting his ideas that entertained the reader. Published in 1859, the book was an immense success and was translated into many foreign languages, including Dutch, German, Danish, Swedish, Spanish, Italian, Turkish, Arabic, Japanese, and native tongues of India. Other books followed - *Character* (1871), *Thrift* (1875), *Duty* (1880), and *Life and Labour* (1887), but none had the success of *Self-help*. On the more technical side of his interests, several volumes were written and reissued under the title of *Lives of Engineers* [7].

His last book was an autobiography [4] that he was encouraged to write. It is clear that Smiles believed in hard work. He mentioned how, after having performed his duties as Editor or as Secretary, he would come home and devote the evening to writing. He was continually expanding his knowledge and collecting information for his writing. One gets the impression of a man who is prepared to work hard, is interested in people of all classes, and constantly trying to improve himself on a wide range of subjects. Thomas Mackay, who edited his autobiography, wrote in his Preface: "Dr. Smiles' achievement is that by common consent he is recognized as the authorized and pious chronicler of the men who founded the industrial greatness of England." A portrait of Smiles is in the National Portrait Gallery in London.

Smiles autobiography does not dwell on his home life. The biography by his granddaughter, Aileen, gives a much more personal account of the life of his family [5]. Smiles had married Sarah Ann Holmes in 1843 and they had three daughters and two sons. By the time they had moved to Blackheath, they moved in similar circles as the Hartrees, though the Hartree circumstances perhaps were more comfortable than those of the Smiles.

There was a certain amount of visiting between the Smiles, Penn, and the Hartree families. In fact, in his *Autobiography* Smiles thanked John Penn for assisting him with information about several eminent engineers.

The eldest of the Smiles' children was Janet, an earnest and conscientious girl, who had been sent to Edinburgh for her education. Having been a dutiful daughter, her parents rewarded her in 1866 with a summer vacation in Switzerland. The group consisted of father, mother, Janet, her sister, and John Penn Hartree or "Jack" as everyone called him, a year younger in age than Janet. He had just completed his studies at Trinity College, Cambridge, obtaining a first-class in the Natural Sciences Tripos. Though Janet considered him a "genius," he was only an inordinately shy and retiring boy as far as she was concerned. By the end of the summer that had changed.

After their return, Dr and Mrs Smiles went to see Mrs Hartree and her children, probably to tell them about their wonderful summer. It was during this visit that the shy, reserved Jack Hartree, who rarely spoke a word, announced his engagement to Janet Smiles. This was a complete surprise to Charlotte Penn Hartree as well. She accused the Smiles of manipulating her son into marriage, knowing that on his twenty-first birthday he was to come into a large fortune. He was not yet 21. This was startling news not only to the Smileses but the prospective groom. Dismayed at the accusations directed against his loved one and her parents, Jack burst into tears on Janet's shoulder [5]. This "family row" has been remembered for several generations.

Of course, marrying for wealth was totally against Smiles' belief. In *Self-help* he had written "riches are oftener an impediment, than a stimulus to action." He did not approve of his daughter marrying a man who had not acquired wealth through his own hard work. But he was a kind man and had no wish to stand in the way of his daughter's happiness. They were married two years later in 1868 and on April 8, 1870, William Hartree, Douglas Hartree's father, was born. Soon after, Jack took up medical studies, first at Cambridge and then Guy's Hospital in London, working for a while in a hospital in London and becoming a Member of the Royal College of Physicians. In 1876, the Jack Hartree family moved to Belfast where some of the Smiles family resided, but they returned to Tonbridge Wells in Kent in order to be near their son's school and here Jack practiced as a physician. All his life he rarely allowed his photograph to be taken.

Samuel Smiles, who showed great interest in the education of his 22 grandchildren had suggested to Janet and Jack that their son William should go to Tonbridge School. William received a very good education at Tonbridge, and according to Aileen, he "won all those scholarships." He was well-prepared for entrance in 1889 to Trinity College, Cambridge, from which he graduated in 1892. After a short apprenticeship, he lectured and demonstrated in the Electrical Engineering Department at Cambridge until 1913. At that point he retired for personal reasons, not needing the income.

Jack Hartree's sister, Isabella Charlotte Hartree, married Edwin Rayner, a prominent physician in Stockport (south-east of Manchester), and at one time Treasurer of the British Medical Association. Their daughter, Eva Rayner, was born on Christmas Eve, 1873. In 1892 she went to Girton College in Cambridge, taking her Tripos in 1895, without being granted a degree since, at that time, Cambridge University did not grant degrees to women. That same year she and her cousin, William Hartree, were married.

Prior to William's retirement Eva did little public work. Besides the birth of Douglas Rayner (1897), she had four more children: Colin William (1898), followed by two children who died in infancy, and John Edwin (1907) who died at the age of 7. Thus even in this comfortable, professional family, with Eva's father a physician of reknown, there was heartbreak. Still she found time to act as honorary secretary of the British Red Cross Society for the Borough, organized and ran first-aid and nursing classes, and was commandant of the Trumpington Veterans Administrative Division.

After William's retirement in 1913, the family took up residence in Farnham, Surrey for a year. There she became secretary of the Farnham Suffrage Society. After another year at Falmouth, they returned to Surrey where she continued in numerous volunteer activities. At this time, their two sons, Douglas and Colin, were attending school nearby.

In 1919, with Douglas about to return to his University studies after WW I, William and Eva returned to Cambridge where they lived for the rest of their lives. Mrs Hartree served on the Borough Council for 20 years and was the first woman Mayor of Cambridge in 1925 (Figure 1.1 shows her in her official gown). Her obituary in the *Cambridge News*, September 10, 1947 lists her many prominent roles, not only local but national and even international. She was an elected member of the Executive Committee of the British National Council of Women, and for a time served as President. For a time she was secretary of the Cambridge branch of the League of Nations Union.

She had a passionate interest in music, an interest that she passed on to her sons, both Douglas and Colin. During the war years, when she had a reason to visit London, she thought not of bombs but of concerts she might have the opportunity to attend. Upon the death of her husband in 1943 she moved to London where there were more concert opportunities, although it also suited her desire to work with refugees.

To achieve all these positions, she had to be a person who was approachable, listened sympathetically and had a quick mind for grasping the facts and assessing the case. She was persistent in her pursuit of what in her heart and mind she thought was right. Her obituary states "Few of us could keep the pace she set, but by alchemy of her understanding personality she kept our love and admiration." Towards the final years of her life "the full sense of the pattern of her hopes being completed came to her when

Fig. 1.1   Eva Rayner Hartree, Mayoress of Cambridge, 1925

it was clear that men and women would have equal rights at Cambridge University; and this pleased her because it was a matter of justice."

William Hartree continued to influence the life of his son, and more will be mentioned in subsequent chapters, but clearly the Hartree family was an exceptional one. In the obituary in honor of William Hartree [8], A V Hill mentions that, in his view, Jack Penn Hartree "passed on his temperament and abilities to his son: retiring, silent, able, kindly, with high standard and perfection of work," qualities that we will see William Hartree, in his turn, passed on to his son. But from this review of the family, we also see a sympathetic but determined mother who no doubt influenced Douglas in many ways.

In many biographies of Douglas Hartree, the only family reference is "great-grandson of the famous social reformer and writer Samuel Smiles, author of the book *Self-help*." In fact, the two had very different personalities. Samuel Smiles "loved to talk" [5], whereas Douglas Hartree was more like his grandfather – shy, not given to small-talk, and never seeking the lime-light. Pictures of him too are rare, usually including only part of the official picture of his Royal Society memoirs. But in the diligent application of his special talents he may have been following Samuel Smiles "self-help" advice.

Notes and references

[1] William Hartree's *Memoirs* were published in 1859 in the *Proceedings of Civil Engineering*. There is some question about the accuracy of these memoirs concerning his education in that Merchant Taylor's School has no record of his attendance at the School, only that he was a Liveryman of the Merchant Taylor's Company.

[2] For further details about John Penn and his son see the article by Prosser R B 1921 *Dictionary of National Biography* Vol. 15, pp 750-751 and references therein, and also *Greenwich Industrial History Newsletter*, 1999, Vol. 2, Issue 5 as well as 2000, Vol. 3, Issue 3

[3] J Davy and Sons 1890 *Catalogue of the Valuable and Extensive Library of the late William Hartree, Esq. of Lewisham* (London: Dryden Press)

[4] Smiles, Samuel 1905 *The autobiography of Samuel Smiles, LLD* Ed. Thomas Mackay (London: John Murray) and (New York: Dutton)

[5] Smiles, Aileen 1956 *Samuel Smiles and his surroundings* (London: R. Hale). This book by Smiles' granddaughter gives an extended family tree of the Smiles family.

[6] Smiles, Samuel 1st Ed. 1859 *Self-help: with illustrations of character and conduct*. There were many later editions, including 1871 *Self-help: with illustrations of character, conduct & perseverance* (New York: Harper). *Self-help* was reprinted, with the addition of an introductory article by Asa Briggs, 1958 (London: John Murray) on the centennial of the original publication.

[7] Smiles, Samuel 1861-62 *Lives of the Engineers* 3 vols. (London: John Murray), and 1863 *Industrial Biography: Iron-Workers and Tool-Makers* (London: John Murray)

[8] Hill A V 1943 *Nature* **152** 154

Chapter 2

Education, World War I, and marriage

Douglas Hartree began his education in Cambridge at St Faith's School. Then in 1910 at the age of 13, together with his 12 year old brother, Colin, he went to Bedales School in Hampshire.

Bedales was one of the early "Progressive Public Schools" [1]. It was founded by John Haden Badley [2; 3], a classicist, educated at Rugby school and Trinity College, Cambridge. He accepted an appointment as assistant master at Abbotsholme, another progressive school, but wanted to get married, something the headmaster would not allow, so he decided to start his own school. He found a site in Sussex, a country house near Lindfield, on Bedales Hill, the origin of the school name. In June 1891, he sketched his ideas in a publication about his aims for a school for boys and its system [4]. The school opened in January 1893 with three students and six staff (three male and three female). The same year he married Amy Garrett whose musical gifts and concern for the emancipation of women were to impact the future of the school. Early Bedales by today's standards was harsh: cold morning baths were the rule, runs before breakfast, caning by prefects, hard labour. But Badley was experimenting and things changed.

It is not exactly clear how Bedales became co-educational in 1899. No doubt Mrs Badley, who was active in the Women's Suffrage Movement, had some influence, but it appears that a mother of one of the boys also had a role. She approached the Badleys with the request that they educate her daughter as well as her son. For two years, the mother with her daughter and three other girls, arranged for them to be day students at Bedales. Then in 1899 the house at Scaynes Hill was provided for girls who attended class and participated in other activities as much as possible. From then on the school was co-educational. In 1900 the school moved to its present site in a tiny village called Steep, near Petersfield, in Hampshire, where the school bought 120 acres. It has remained there ever since. By 1914, when Hartree was there, it had 100 boys, 60 girls in the main school, and 40 pupils in the junior house, with about 20 teachers.

The school motto is "Work of Each for Weal of All." Its guiding principle is sustained by a set of values which emphasizes the importance of individual development though keeping it clearly related to the community and society in which the individual lives, looking at that society in its widest sense. The conventional, strongly church based and often imperialistic values of the normal English public school of the time were largely rejected.

It is likely that the choice of school was Eva Hartree's since, in her work with women's causes, she must have known or even met Mrs Badley. But the school's strong emphasis on academics and music, and a relatively lesser emphasis on team sports, also matched Hartree's interests. A few years later, Elaine Charlton, the future Mrs Douglas Hartree, also attended Bedales.

The *Bedales Record*, a school year book, notes that Douglas took an active part in the school's activities. After only one year he was "running the meteorological office." The following year (1912-13) he presented a required contribution to the Senior Debate, a lecture to the Musical Society on *The theory of sound and its relation to music*, and was elected to the committee of the Scientific Society. The latter printed a magazine in which Hartree published his first paper, on *Observations on certain periodic properties of Numbers*. In his final year, Douglas, together with his brother Colin and three others, constructed a wireless station. He was now on the Music Committee, secretary of the Scientific Society and, in one of his last school activities, he played the part of a policeman in "Pirates of Penzance."

The *Bedales Chronicle* was a magazine run by the pupils, with several issues a year. Hartree's first contribution was an article on *The May races, Cambridge, 1913* printed in the June 1913 issue. The style is that of a reporter and the article starts with:

> June 4. It is a fine day, and the first day of the "Mays," so let us take some conveyance – or walk – and see whether we can get any good spot to watch the races.

He then explains how the races are conducted, the process of "bumping." He describes the boats getting ready, the excitement and noise of the crowd, the race itself. He ends with "our party breaks up, and we go home, well satisfied with the first day at the 'Mays'." It could only be written by someone well-entrenched in Cambridge lore.

Hartree's close friend at Bedales was Lancelot Law Whyte. Both were vying for scholarships at Cambridge and the whole school took an interest in their progress. The December 1913 issue of the *Bedales Chronicle* mentions that they had been up to Cambridge for scholarship examinations but it had really only been a trial attempt before the more serious attempt next year. A year later it is reported that "In December, L L Whyte and D R Hartree went up for their scholarship examinations at Cambridge and fulfilled our highest hopes, for both obtained the highest scholarships awarded: Whyte at Trinity and Hartree at St John's." A whole day holiday was promised to the school in honor of their success to be held in the summer when it could be better spent. It was held on May 25, 1915. On leaving Bedales, Hartree presented the book *Railways of the World* to the school library. We will see that many of his interests started at Bedales.

During his first year at Cambridge, Hartree read for the Mathematical Tripos. This choice, rather than the Natural Sciences Tripos, was often recommended by Cambridge tutors for students who were theoretically inclined. Both pure mathematics and applied mathematics were covered in the curriculum. He obtained First Class Honors in the examinations at the end of the academic year in 1916. At that point he took leave from Cambridge to participate in war work.

World War I and its aftermath

In order to keep the troops in France supplied with shells and ammunition, the British government created a Ministry of Munitions. It also was given the task of developing new weapons. Air raids on England early in the war led to the need for defense against aircraft. The Munitions Inventions Department of the Ministry of Munitions created an Anti-Aircraft Experimental Section for this purpose.

A V Hill (1886 - 1977), from Trinity College, Cambridge, was asked to form and direct the Section [5]. Lieutenant R H Fowler (1889 - 1944), a Royal Marine artillery gunnery officer, was recommended to him and soon was promoted to Assistant Director. Fowler too was educated at Trinity College having graduated in 1911. They were joined by Edward A Milne (1896 - 1950) who, according to Hill [6], was "an essential element in our party and did some of the finest work we achieved."

In January 1916, Hill was asked to investigate some urgent problems in anti-aircraft gunnery. The three – Hill, Fowler, Milne – started field tests at Northolt aerodrome in Middlesex, where they were joined by William Hartree and A C Hawkins. The party moved to the National Physical Laboratory in May 1916, and were soon joined by D R Hartree and H W Richmond, a Fellow of the Royal Society (FRS) in mathematics.

In August 1916 the whole party was moved to HMS Excellent, Whale Island, Portsmouth, where they remained for the rest of the war. Fowler resided in Portsmouth, while Hill spent most of his time in London staying in contact with service needs. Most participants in this work were commissioned. Hartree was a Lieutenant in the Royal Naval Volunteer Reserve and in Figure 2.1 is shown in his uniform. When gun-trials and fuse-trials became extensive many others joined the group at Portsmouth, including Colin William Hartree. In the Services the group was known as "Hill's Brigands." According to Milne [7], it was tremendous "fun" collaborating with Fowler. Usually it was not he who produced the first idea; "but he tossed it back at lightening speed ..." There was much routine computing as well as actual mathematical research in ballistics. The high-angle trials of anti-aircraft shells and fuses were observed from Eastney or from Hayling Island. All shared in the tasks at hand, even if that meant a disproportionate effort was needed on the part of Richmond, well over fifty, for cycling over to Eastney or Hayling Island. The experience of watching Hill, Fowler, and Milne direct research and be part of a team with Richmond, was "far better training than most Universities could offer to aspirants to research." The latter was said by Milne, writing about Hill and Fowler, but it seems appropriate to include Milne also.

World War I was officially over when the armistice was signed on November 11, 1918. Hill and Fowler returned to Cambridge in the spring of 1919, and other Cambridge members of the group returned when they were demobilized.

Before committing his time to studies, Hartree wrote the first of many short articles he was to publish in *Nature* during his life, this one on ballistic calculations. It is a clearly written outline of numerical methods developed during the war for describing the motion of a shell after leaving a gun. He divided the problems into two groups, primary and secondary. The primary problems were concerned with the calculation of the motion of a shell fired under ideal conditions, such as still air and a standard muzzle velocity. It was assumed that still air gave a resistance in a direction opposite to the motion of the shell. Earth rotation was neglected so the shell trajectories

Fig. 2.1 Lieutenant Douglas R Hartree of the Royal Naval Volunteer Reserve (about 1916).

were in the vertical plane containing the initial direction of motion. Secondary problems included the effects of the rotation of the Earth, together with the winds and changes in the atmospheric density, both of which might vary along the trajectory. Before the advent of anti-aircraft gunnery, the point of impact on the ground was the only point of practical importance, as for example, when firing at fortifications. Now it was necessary to compute whole trajectories. The method of approach was to solve for a plane trajectory by step-by-step integration for selected cases and interpolate to obtain intermediate trajectories. In some instances, smoothing was necessary. Secondary problems were made manageable by computing "first order" effects which were assumed to be additive. In the course of this work he learned, in a very practical way, about numerical integration of systems of differential equations, interpolation, smoothing, and equations of variation for non-linear problems. The majority of the work was carried out by means of hand operated calculating machines. The paper appeared in 1920. In later years, he pointed out that prior to 1916 practical knowledge of anti-aircraft gunnery was limited to grouse-shooting.

This work on the integration of differential equations for trajectories had considerable impact as explained by Darwin in later years [8].

> ...at the age of 20 he introduced a reform which may not now seem very revolutionary, but which must have been rather disturbing for the older practitioners of the science. By tradition the angle of elevation of the trajectory had always been the independent variable, and the co-ordinates and time were expressed in terms of it. He showed that there would be great advantages in using time itself as the independent variable, and expressing all other quantities in terms of it. He introduced this practice and it is now generally accepted.

In the Fall of 1919, Hartree resumed his studies at Cambridge, choosing to take the Natural Sciences Tripos. Students who had passed the Mathematical Tripos Part I and had not taken Part I of the Natural Sciences Tripos, were expected to take at least two years for Part II in one subject, physics in Hartree's case.

Early in 1920, the family suffered a devastating tragedy. While Douglas had been at Bedales, his seven year old brother John Edwin had died. Douglas had gotten to know him well, but he was much closer to his brother Colin. When Colin started school at Bedales, he suffered from a delicate constitution, but his health seemed to steadily improve. Like Douglas, he was fond of music and a clear thinker, but in contrast to Douglas, was a good football player. Then on February 9, 1920, Colin died of meningitis at the age of 22. This was extremely hard on all the family. For Hartree, the infant deaths of a brother and sister when he was no more than 7 years of age, followed by the loss of a brother he had gotten to know well when he was 17, and another when he was 23, leaving him the only child of Eva and William Hartree, was a particularly painful experience. He was known to become emotionally moved when talking about John and Colin.

In the final examination in 1922 Hartree obtained a Second Class, possibly because of the disruption to his education by the war, as is usually stated, but the death of his brother and the toll that took on the family must also have been a factor. This completed his undergraduate education and he received his BA degree. Figure 2.2 is a portrait of Hartree at about this time.

In the meantime, Hartree's friend L L Whyte had gone to the Somme, where he served as a Lieutenant in the Artillery obtaining a Military Cross, but also becoming seriously wounded. After the Armistice, he returned to Trinity completing his undergraduate studies, then continued research under Rutherford, spending some time in Göttingen with Born and returning to Cambridge to work with Hartree and his friend, Patrick Blackett, at the Cavendish. But the Somme still haunted him and he left in 1923 for a career as "philosopher, scientist, banker." He wrote a number of books on topics ranging from *The Atomic Problem* to *The Unconscious before Freud*. His obituary appeared in the *Times*, having died suddenly on September 14, 1972.

For William Hartree too, the end of the war and the family tragedy required adjustments. When World War I started in 1914, William Hartree had volunteered to work with the Post Office as a telegraph linesman. At the suggestion of his uncle, Edwin H Rayner, A V Hill invited him in 1916 to join the newly founded Anti-Aircraft Section. In William Hartree's obituary, A V Hill wrote [9]

> It was hard to believe that this shabby, middle-aged linesman was an able engineer and mathematician – but he was, and much more. Whenever a job of hard work had to be done on time, whenever something needed to be fetched or carried, whenever long hours and discomfort had to be to be endured "at the far end of the base," Hartree was there. Nobody could see shell-bursts so nearly into the sun, nobody could record what he saw so accurately and quickly, nobody could interpret the results so well, nobody could come to the office so early or stay so late to work them out. Quietly one day he improvised a long-base height-finder out of some wires, posts, and a steel tape.

Fig. 2.2 Douglas Hartree at about the time of his graduation, 1922.

Fig. 2.3   William Hartree with a spiral (cylindrical) and linear slide rule.

It came to be called the Hartree height-finder and was used extensively by the troops until better optical height-finders were produced. It is believed that it was for this that he was given an OBE (Order of the British Empire). On Christmas Day, 1918, he sent A V Hill four volumes of his grandfather's book, Lives of the Engineers, with a note thanking Hill for his kindness. He later approached Hill saying he had never enjoyed anything as much and wished he could continue similar work. Hill invited him to join him at Cambridge and do physiology.

For the next thirteen years, they worked together on the physiology of muscle, its heat-production, and various aspects of its dynamics. William Hartree's first scientific paper appeared in 1920 when he was fifty. He was known, except for his work, by very few: he never attended meetings. Hill claimed that "In a sense he was an amateur who worked for the love of his work with the intensity and pride of a craftsman." Hill said that in their work together, Hartree believed he had written down between 10^7 to 10^8 figures and very seldom were mistakes discovered. He published 17 papers jointly with Hill, 5 papers jointly with other authors, and at least 10 papers in his own name, a remarkable output for someone who started research at the age of 49. The picture in Figure 2.3 shows W Hartree with both a spiral (cylindrical) and a linear slide rule that he used frequently in his research.

Though William Hartree continued to work in Cambridge, Hill himself left in 1920 to become Brackenburg Professor of Physiology at Manchester University where he remained until 1923. In 1922 Archibald Vivian Hill was awarded the Nobel Prize in physiology or medicine "for his discovery relating to the production of heat in the muscle," work to which W Hartree had contributed.

The enjoyment and pride Douglas saw his father take in research must have influenced the son, since it was not too many years later when he too decided to go into research.

Reunions at Bedales and marriage

In 1906, the first meeting of Old Bedalians (OB) was held at the school and became an annual event thereafter. Douglas Hartree maintained his tie to Bedales. In 1916, one year after leaving Bedales, he published a contribution to the discussion on *The present upheavals in art and literature are signs of decadence rather than progress* in the *Bedales Magazine*. He responded with the comment that many developments in music, especially harmonies, had precursors and used examples from Bach to make his point. He wrote from first hand musical experience, since radio and recordings did not exist at that time.

At one of these reunions, he noticed Elaine Charlton, also an OB. They had overlapped two years (1913-15), though they scarcely met at the time. After leaving Bedales, Elaine had gone on to study piano with Tobias Matthay at London, one of the most distinguished piano teachers during his time, teaching by his own method at the Royal Academy of Music. It appears that Bedales and music brought them together and they began seeing more and more of each other. In 1923 they were married, with Patrick Blackett, then a research student under Rutherford at the Cavendish Laboratory, as Hartree's best man. Before their marriage, Hartree had informed Elaine that he was determined to go into research and that sometimes inevitably it would come first [10]. For Elaine, it was hard at times to understand the creative mind, particularly when Hartree said he had a "brain wave" and was not available for the family. But Bedales continued to be important in their lives: their daughter Margaret (born 1925) was there in 1937-40, and their son (John) Richard (born 1931) in 1944-50, becoming Head Boy. Only their first son Oliver (born 1927) did

Fig. 2.4 The wedding party of Douglas and Elaine Hartree. Left to right, front row: Elaine and Douglas, male figures behind are William Hartree (father of the groom), Patrick Blackett (best man), and Eustace Charlton (father of the bride).

not attend. Pictures of Douglas and Elaine's wedding, Douglas with the Charlton family, and a family picnic are shown in Figure 2.4, Figure 2.5, and Figure 2.6, respectively.

During his entire life, Hartree seemed to be comfortable working with women on an equal basis. Possibly this too may be attributed in part to Bedales, though seeing his mother in her many activities must also have been a factor. In 1925, when Bertha Swirles began her research studies under Fowler, Hartree generously passed on unsolved problems from his research to her. In July 1925 she had occasion to go to their house, Latrigg, named after a hill near Elaine's home in Keswick. Bertha wrote [11]:

> At the home of Douglas and Elaine, I recognized at once a common interest in music as the room contained two pianos — Elaine's Steinway and an upright. She was an accomplished pianist, and Douglas had a deep and wide knowledge of orchestra and chamber music. On this occasion he lent me his notes of Fowler's lectures of the previous term and two chapters of the dissertation for the PhD on which he was then working.

We will see that they continued to collaborate.

The house on 21 Bentley Road had been specially built for them, a gift from Douglas' parents and less than a mile from where they lived. It had four bedrooms, a bathroom, and study upstairs, and a drawing room with two pianos, dining room, kitchen, pantry, and maid's room downstairs. The 7' 6" Steinway grand had been a 21st birthday present from Elaine's parents. There also was a large garden with a lawn for a full sized tennis court. Both Douglas and Elaine played. After baby Margaret arrived in 1925, a nanny would live in.

Notes and References

[1] Public schools in Britain, such as Eton, Rugby, or Tonbridge, were elite boarding schools, strongly church based, that by the 1800's had become a means of upward mobility from the upper-middle class to aristocracy. "Progressive" public schools rejected these values. They were non-sectarian, with less of an emphasis on classics and organized games, and more emphasis on science and modern languages, an education that recognized the importance of science and technology.

[2] Badley J H 1923 *Bedales - A Pioneer School* (London: Methuen)

[3] Henderson J L 1978 *Irregularly Bold. A Study of Bedales School* (London: Andre Deutsch) pp 1-42

[4] Badley J H 1892 *Bedales (Hayworths Heath, Sussex). A School for Boys.* (Cambridge University Press)

[5] Katz B 1978 *Biographical Memoirs of Fellows of the Royal Society* **24** 71

[6] McCrea W H 1951 *Mon. Not. Roy. Astron. Soc.* **111** 160

[7] Milne E A 1945 *Mon. Not. Roy. Astron. Soc.* **105** 80

[8] Darwin C G 1958 *Biographical Memoirs of Fellows of the Royal Society* **4** 103

[9] Hill A V 1943 *Nature* **152** 154

[10] Notes of an interview with Elaine Hartree, conducted by T S Kuhn on May 12, 1963, Center for History of Physics, American Institute of Physics, College Park, MD

[11] Jeffreys, Bertha Swirles 1987 *Comments At. Mol. Phys.* **20** 189

Fig. 2.5   Douglas Hartree with the Charlton family in 1919. Elaine is at the lower right in the picture, next to her younger brothers.

Fig. 2.6 A family picnic (June, 1923). Left to right: William and Eva Hartree, unknown, Elaine and Douglas Hartree.

Chapter 3

Early research at Cambridge University

The 1920's were exciting times in atomic physics. In 1921, Albert Einstein was awarded the Nobel Prize "for his services to Theoretical Physics, and especially for the discovery of the law of the photoelectric effect." That was also the year Niels Bohr visited Cambridge to give a course of lectures on quantum theory. Hartree later considered this visit by Bohr the most important event in his scientific career [1].

Hartree entered the PhD program in the Fall of 1922, intending to do research on controlled excitation of spectrum lines by electron bombardment in the Cavendish Laboratory where Ernest Rutherford was the Professor of Physics.

Prior to his professorship at the Cavendish, Rutherford had been Langworthy Professor of Physics at the University of Manchester. In 1910 he began investigations into the inner structure of the atom that led to the concept of the "nucleus," his greatest contribution to physics. He discovered that the atom was not a hard sphere, but mostly empty space, with a nucleus surrounded by electrons. In the Spring of 1912, Niels Bohr joined Rutherford at Manchester and, by applying Max Planck's quantum theory to Rutherford's atom, obtained a theory of atomic structure based on quantized orbits. This led to a view of atoms much like a miniature solar system but with an important difference: the energy of an atom could only be one of a certain set of discrete energies.

In 1919 Rutherford accepted a position at Cambridge as Cavendish Professor. There he was an inspiring leader. He steered many future Nobel Prize winners towards their achievements while other laureates worked with him at the Cavendish. Thus it was a flourishing research environment in nuclear physics. During that time, "physics" essentially meant "experimental physics" so those who were theoretically inclined were on their own to a large extent. For theoretical physics students, there was nowhere to work except a rather small library which also served as a tea room [2].

After the war, R H Fowler had been appointed Lecturer in Mathematics and Fellow of Trinity College. A friendship sprang up between Fowler and Rutherford that extended to all of Rutherford's students and co-workers. In fact, many of the theory students in the Cavendish, after Hartree, were advised by Fowler who, in 1921, married Eileen, Rutherford's only daughter. Thus there were several strong bonds between Rutherford and Fowler.

Hartree soon discovered he had no aptitude for experimental physics that required the building of an apparatus. Remembering Bohr's inspiring lectures, he looked for problems in quantum theory. For this research, he found a mentor in R H Fowler with whom he had conducted wartime research as a member of the "Brigands." In fact, they were co-authors of a paper, *The pressure distribution on the head of a shell moving at high velocities* that had been published in 1920.

After Bohr's lectures at Cambridge, Fowler had noted [3]:

> Since Bohr's generalized theory of spectra and atomic constitution was published in 1922, there was a great need and opportunity for a more quantitative application of it than any yet published by the author. Such an application must involve much numerical computation and is therefore likely to appear unattractive to anyone not well trained in such work, who must at the same time possess an intimate knowledge of modern physical theory.

Hartree seized the opportunity entirely on his own initiative.

Though Fowler is often mentioned as Hartree's supervisor, a list of Fowler's students [4] compiled from published University records, does not include D R Hartree. Since Hartree was a member of the Cavendish rather than the Mathematics Department, it appears he was automatically enrolled as Rutherford's student. However comments on his dissertation were written by Fowler [3].

Hartree's first paper on atomic spectroscopy was based on Bohr's theory, which had been extended by Sommerfeld to include non-circular orbits. In this theory, electrons moved in orbits in a plane, characterized by their minimum and maximum distance from the nucleus. He studied systems where there was only one electron in the most lightly bound orbit. A fundamental concept was the fact that light, or radiation in general, consisted of light quanta (photons) of energy $h\nu$, h being Planck's constant and ν the frequency. When atoms were "struck" by "photons" nothing would happen *unless* the energy of the photon was just right to "excite" the atom to another state or large enough to remove the electron. Thus there were specific wavelengths for excitations and excited states could decay, emitting photons corresponding to the difference in energy of the two states. This difference defined the wavelength or, alternately, the wave number of the emitted photons.

Starting from fundamental principals and Bohr's conditions, Hartree derived an equation for the orbit which, of course, involved many physical constants. In his typical *modus operandi* he redefined variables (or units) so as to obtain the uncluttered equation

$$n - k = \oint \left[-\frac{\nu}{R} + 2v - \frac{k^2}{\rho^2} \right]^{1/2} d\rho, \qquad (3.1)$$

where the integers n and k are principal and orbital quantum numbers, respectively, R is the Rydberg constant, ν the wave number of the orbit, and v is a potential function derived from the field. There could be two possible formulations of a problem based on this equation: given the field of force, determine the term energy (ν/R) of the orbits *or* given the observed term energies, find the field of force. Hartree chose the latter, a semi-empirical method making use of observed data. Both optical (valence) and X-ray (core) orbits were considered along with the assumption that at the same point in space the field was the same for all types of orbits.

There was a difficulty in dealing numerically with the function v in that it ranged in magnitude from zero to infinity and its use could lead to computational problems. So Hartree introduced another equation involving the notion of an "effective nuclear charge," $Z(r)$, that varied smoothly from the nuclear charge, which he denoted by N, and the net core charge C at large distances. The two functions were related through the equation

$$v = -\int^\infty \frac{Z}{r^2} dr = \int_0 Z d\left(\frac{1}{r}\right). \qquad (3.2)$$

This greatly simplified the work even though an extra integration was needed. With this rearrangement it was possible for him to arrange the computational part in such a way that *no apparatus was required beyond a 10-inch slide rule and a table of squares*, along with an adequate amount of paper for recording intermediate results.

For a given ν/R and k, and an estimated function $Z(r)$, he obtained a computed n, denoted by n_c, which should be close to an integer. To validate Bohr's theory, he made use of the notion of a "quantum defect"

(q) defined by the relation

$$\nu/R = C^2/(n-q)^2. \tag{3.3}$$

Then $\delta = n_c - n$ should be close to q, all being consistent. Agreement varied. In the case of sodium, he also attempted to relate the scattering power of a single atom, computed from his orbits, with similar information observed by Bragg et al [5] in the study of X-ray reflection in rock-salt crystals. In this first paper, read to the Cambridge Philosophical Society on May 21, 1923, Hartree applied these methods to the study of potassium, ionized calcium, and sodium with modest success. This paper is important, not for his results (the theory on which it was based was inadequate), but as a demonstration of his insight into how a problem can be defined and reformulated in such a way that its solution is more amenable to computation and how he tried to apply his results to problems of current interest.

In another paper, read to the Cambridge Philosophical Society on May 19, 1924, he described his attempts to identify lines of homologous series spectra (spectra of atoms or ions with different nuclear charges but the same electronic structure outside filled shells) from among the many lines observed by taking advantage of regularities and interpolation. He introduced the terms "lithium-like" and "sodium-like," terms still used today, pointing out that Millikan and Bowen [6] used the term "hydrogen-like" for any atom with one valence electron. Following the notation of A Fowler and Paschen [7], he used the core charge C as a spectrum number in Roman numerals so that the doubly ionized aluminum was named Al III. His analysis was based on Equation (3.3) and the approximate relation:

$$\nu/R = C^2/n^2 + \alpha \tag{3.4}$$

where α was the "term excess." In Bohr's theory, two types of orbits were considered: i) circular orbits that remain well outside the core and ii) orbits which penetrate the core. He felt that one relation might be more regular along a series than the other. By devising relationships that were nearly linear with nuclear charge N, unknown term values could be interpolated or extrapolated from known values and lines in spectra identified. According to an addendum to the proof, some of his identifications were confirmed experimentally.

A revised and extended version of these two papers by Hartree, was submitted by Rutherford to the Proceedings of the Royal Society. In this paper, Hartree related the "term excess" for the circular orbits to polarization of the series electron. For orbits that penetrate the core, he found that $1/q$ was approximately linear in the atomic number.

In the early 1920's, Fowler was investigating the properties of gases at very high temperatures, such as those found at the center of a star. For this work, it was necessary to know the successive ionization potentials of each atom being considered, where the ionization potential is the energy required to remove an electron. He asked Hartree to investigate this process and so he did, reporting all the successive potentials of oxygen, Fe XXVI - Fe XVI, and all for silver (N=47).

In one of the most famous papers on atomic spectroscopic analysis, Catalán in 1922 [8] had discovered multiplets in Mn I and Mn II spectra. This paper stimulated the interpretation of the structure of complex spectra. Hartree showed that one could make estimates of ionization potentials from the wavelength of a resonance line and the difference of quantum defects for the two levels. Using this method, he calculated the ionization potential for the removal of a $4s$ electron from the $3d^54s$ ground state of Mn II. He obtained 14.5 eV: this may be compared with the modern value of 15.6 eV. Hartree's estimate was remarkably good given the primitive state of spectroscopic analysis of complex spectra at the time.

As a research student, Hartree regularly attended the Kapitza Club (for research students in the Cavendish Laboratory), the $\nabla^2 V$ Club (an informal club for the discussion of ideas in theoretical physics), and the Cambridge Philosophical Society. All of these organizations provided opportunities for students to be exposed to new research ideas and to develop skills for presenting their work.

Any Fellow of the Cambridge Philosophical Society could communicate papers for publication, whereas others had to get a Fellow to communicate their papers for them. In the case of a student, this often was done by the Fellow who had been closely involved with the research. Only Cambridge graduates could become Fellows but had to be nominated by three other Fellows. Hartree became a Fellow of the Cambridge Philosophical Society in 1923, one year after having received his first degree.

After his first few publications demonstrating his research ability, Hartree became a Fellow of St John's College for the period 1924-1927.

Fig. 3.1 The class of Paul Ehrenfest (near the center) on the occasion of a visit by Albert Einstein and Douglas Hartree (between Ehrenfest and Einstein). (See the article by Samuel Goudsmit, who appears on the right, in *Physics Today* June, 1976 for a complete class list. The picture is reprinted by permission of the Center for History of Physics, American Institute of Physics, College Park, MD)

Passing through Leiden

It is not known whether Hartree studied in continental Europe where many advances in quantum theory were occurring. We do know that his friend, L L Whyte, spent time in Göttingen. We also know (indirectly) that in November of 1923, Douglas Hartree was passing through Leiden visiting Paul Ehrenfest who was teaching at the Institute for Theoretical Physics at the University of Leiden. Figure 3.1 shows a picture of Douglas Hartree with Professor Ehrenfest's entire class. It happened that Einstein too was visiting, having fled to Leiden after Hitler's "Beer Hall Putsch" in early November, fearing threats on his life should a Nazi regime came to power [9]. Thus the picture was taken of Hartree between Ehrenfest and Einstein. Only half of the students were physicists, the rest were astronomers and chemists [10]. It is likely that Hartree presented his research during this visit. In Hartree's first paper he acknowledged indebtedness to Professor

Ehrenfest for a more convincing reason for adopting a procedure he had decided to use.

During the visit, Hartree met Samuel Goudsmit (1902-1978), a Dutch student of Ehrenfest's who in 1925, together with another student, George Uhlenbeck (1900-1988), put forward the proposal of electron spin. This was a concept of profound significance to quantum theory in that it explained fine-structure splittings in spectra. On March 15, 1924 Hartree sent a letter to Goudsmit, saying:

> Briefly, I have been able to obtain approximate theoretical relations between the terms of the optical spectra of atoms of the same electron structure but of different charge ..., and I have been using these relations to help in the identification of various spectra among the lines of the hot spark spectra of different elements given by Milliken and Bowen in the January, 1924 number of the Physical Review.

He included his results for Li-like systems, and mentioned that he felt certain that the pair Milliken and Bowen reported for C IV were wrong. He also sent his identifications of the fine-structure splitting of the Na-like sequence up to S VI. These certainly must have interested Goudsmit.

What is amazing is that two and a half months after the publication by Milliken and Bowen appeared in the American journal, Hartree had already been working on the identification of lines for two months. He must have been extremely conscientious about being current in his reading of research journals.

Robert Bruce Lindsay and his concept of self-consistency

In the early 1920's, Robert Bruce Lindsay was working on similar problems.

Lindsay started graduate study at the Masssachusetts Institute of Technology (MIT) in the fall of 1920 and took a course which, among other things, included Sommerfeld's application of the Hamiltonian-Jacobian version of advanced quantum mechanics to elliptical orbits of the electrons in Bohr theory. He was encouraged to apply for a fellowship from the American-Scandinavian Society for study at the Bohr Institute in Copenhagen. In this he was successful and was granted a stipend of $1,000 for the year 1922-23 for study of atomic theory with Bohr.

A few years before, Bohr had achieved a scientific breakthrough in his endeavors to understand the periodic system of elements, but had laboured in vain to evaluate the details of a "polyelectronic atom" as he called it.

Even for helium he could not get his method to work. Towards the end of World War I, he had decided to approach the problem empirically and qualitatively. He was trying to fit experimental data from optical and X-ray spectra and also collision results into a systematic pattern. The result was the periodic system of elements, in which elements of increasing atomic weight (number) were considered built up by the capture of electrons by the appropriate nucleus. Detailed calculations were at first passed up but the scheme was sufficiently plausible for him to predict properties of certain elements. He predicted element N=72 (Hafnium, meaning harbor or "haven" in Danish, in honor of Copenhagen's harbor). He received the Nobel Prize in the winter of 1922-3.

On Lindsay's arrival in Copenhagen in the Fall of 1922, Bohr suggested research on some approximations to the orbits of extranuclear electrons for a homologous series. It would involve quantizing the energy states for the various electrons in the field of both the nucleus and the other electrons. A good deal of numerical integration would be involved. It was some time before Lindsay caught on to effective quantum numbers (the $n - q$ introduced earlier). A penetrating orbit spent part of its time inside the inner shells of the atom and hence was exposed to a large part of the unscreened field of the nucleus.

The charge distribution of an electron, as determined from its orbit, is related to the potential ($v(r)$ in Equation (3.1)) which in turn depends on the charge distribution of the other electrons. The negative charge of core electrons and orbits that penetrate the core, from the point of view of an outer electron, shield the positively charged nucleus. As a result of extensive reading there dawned on Lindsay the idea of representing the screening of the inner electrons on the outer ones in terms of a continuous negative charge distribution. From this, there arose the concept of self-consistency, which as far as he was aware had not been suggested before. His scheme, however crude, made no use of empirical data. Note also, that in Bohr's theory there was no differential equation, only an integration. Lindsay spent much of his time [11]:

> ... grinding away at calculations. These for the most part were reduced to numerical integrations, which in turn I did by *counting squares on graph paper*. I certainly learned with a vengeance the real meaning of integration! Much repetition of the same calculation was needed in order to meet the the consistency requirement, and the whole thing became monotonous at times.

Returning to the US, Lindsay was awarded his PhD in June 1924. Soon after he accepted a position at Brown University and started a research career in acoustical theory.

As Lindsay noted in his autobiography, "Ultimately, Hartree applied the scheme to the quantum mechanical evaluation of atomic fields under the title 'SCF method.' In one of the early papers, he gave me credit for making calculations based fundamentally on this idea. Others too, notably Slater, mentioned my contribution." Thus the idea of a "self-consistency" predates Hartree's work to be described in the next chapter. There we will see that the two finally met and collaborated for quite a few years.

It is interesting to note that, in 1923, Lindsay and his wife translated Kramer's (a student of Bohr's) book from Danish to English. The latter was entitled *The Atom and the Bohr theory of its structure*. Similarly, in 1927 Hartree published a translation of Born's *Atommechanik* from German to English, entitling it *The mechanics of the atom*. In this he was assisted by his wife, Elaine, whose German was considerably better than Hartree's.

Notes and references

[1] Darwin C G 1958 *Biographical Memoirs of Fellows of the Royal Society* **4** 103
[2] Mott, Sir Nevill Francis 1987 *A Life in Science* (London: Taylor & Francis)
[3] Bertha Swirles Jeffreys wrote a personal account of Hartree's work which was published in 1987 *Comments At. Mol. Phys.* **20** 189
[4] A copy of an annual list of PhD students registered at Cambridge University with R H Fowler as supervisor, prepared by W H McCrea, was sent by B Jeffreys to R Garstang, who provided me with a copy (2000).
[5] Bragg W L, James R W, and Bosanquet C H 1921 *Phil. Mag.* **41** 309; ibid 1922 **43** 429
[6] Millikan R A and Bowen I C 1924 *Phys. Rev.* **22** 1
[7] Fowler A 1923 *Proc. Roy. Soc.* **103** 413 and Paschen F 1923 *Annal. der Physik* **71** 142
[8] Catalán M A *Phil. Trans. Roy. Soc.* **223A** 127
[9] Medawar J and Pyke D 2001 *Hitler's Gift* (New York: Arcade Publishing)
[10] Goudsmit S A 1976 *Physics Today* June, pp 40. Einstein and Ehrenfest were great friends and, according to Goudsmit, Einstein visited Leiden once a year.
[11] Lindsay, Robert Bruce ca. 1983 *Intellectual autobiography*. This autobiography was made available by permission of the Niels Bohr Library, Center for History of Physics, at the American Institute of Physics.

Chapter 4

The new quantum mechanics

In the meantime, a new quantum mechanics was being born. In the summer of 1925 Heisenberg visited Cambridge, presenting his ideas for "matrix mechanics." Then in 1926, Schrödinger formulated his "wave mechanics" and Dirac gave his first course of lectures, *Quantum Mechanics (Recent Developments)*. Hartree continued his work on orbit theory, submitting his dissertation for the PhD on June 17, 1926. Nearly all the work had been done before Schrödinger's breakthrough, and the introduction in his dissertation included the somewhat disheartened remark:

> .. it seems probable that in the course of time the whole of the quantum theory may have to be reformulated in terms of the new mechanics and a reformulation will supersede most if not quite all the work here considered.

The degree was awarded since it provided an excellent jumping off point for a treatment based on Schrödinger's theory where charge distribution was described by a differential equation.

Hartree lost no time in developing his most important papers in atomic theory, entitled *The wave mechanics of an atom with a non-Coulomb central field*.

Part I Theory and Methods,
Part II Results and discussion,
Part III Term values and intensities in series in optical spectra,
Part IV Further results relating to terms of the optical spectrum.

Part I and II were both read on November 21, 1927 at the Cambridge Philosophical Society. It must have been an interesting event.

At the beginning of Part I, Hartree emphasized a point of contrast between the old orbit theory and the new wave mechanics. In the former, the motion of the electron is described by an orbit in a plane that lies wholly between two radii, whereas in the latter the wave function in all of space is different from zero at all but a finite number of values of the radius, and depends on the field at all distances. Thus there could be no separation into "penetrating" and "non-penetrating" orbits.

This raised the question of assignment of quantum numbers. Hartree suggested that $l = k - 1$, and that n be such that $n - l + 1$ be the number of values of the radius r for which the wave function $\Psi = 0$. This agreed with Bohr's principal quantum number for the hydrogen atom.

As in Bohr's theory, Hartree's first goal was to express Schrödinger's equation in as simple a form as possible, eliminating the various physical constants. This he did through defining the so-called "atomic units:"

Unit of length	$a_H = h^2/4\pi^2 me^2$
Unit of charge	e, the magnitude of charge on an electron
Unit of mass	m, the mass of the electron
Unit of action	$h/2\pi$
Unit of energy	$e^2/a = 2hcR$, twice the ionization energy of the hydrogen atom with fixed (infinite) nucleus
Unit of time	$1/4\pi cR$

The unit of energy today is called a "Hartree" and denoted by E_h. In these units, the wave equation for the motion in space of a point electron with total energy E, in a static field in which its potential energy is V, is defined by

$$\nabla^2 \Psi + 2(E - V)\Psi = 0. \tag{4.1}$$

If the field is of spherical symmetry so that $V = -v(r)$, in spherical polar co-ordinates, Ψ separates into a known spherical harmonic function, $S(\theta, \phi)$, and a radial function, $\chi(r) = P(r)/r$. Then $P(r)$ satisfies the differential equation

$$d^2 P/dr^2 + \left[2v - \epsilon - l(l+1)/r^2\right] P = 0, \tag{4.2}$$

where $\epsilon = -2E$ and is related to the wave number through $\epsilon = \nu/R$. He was careful to point out the advantage of working with $P(r)$ rather than $\chi(r)$, the first being that the differential equation was simpler, and the second that $P^2(r)$ (suitably normalized) had the physical interpretation of being a charge density at radius r.

Hartree then went on to discuss the properties of the solution, in the region of large r, small r, variation of the solution (as when v changes by a small amount), and when v differs negligibly from the Coulomb potential. He ended with discussing numerical methods for the solution of the radial equation.

In Part II, Hartree again mentioned the two possible approaches for numerical work:

(1) a semi-empirical approach for finding the field of force of an atom that reproduces the observed terms as closely as possible, or
(2) finding a field of force such that the distribution of charge given by the wave functions for the core electrons shall reproduce the field.

In his earlier work he had chosen the former, but now, like Lindsay before him, he chose the latter, a totally theoretical method, not relying on observation.

Some assumptions would be needed. Hartree considered the neutral atom of an alkali metal consisting of closed groups (or shells now denoted by nl quantum numbers) and one outer electron. Closed groups have the property of being spherically symmetric and hence the potential for the outer electron had the desired property of moving in a symmetric field defined by the core. But how was the field for the core electrons to be determined? One could remove one electron from the core and say the field for this electron was defined in terms of the field of the remaining electrons, but Hartree knew that if you removed an electron from a filled shell, thereby creating a hole, that the spherical symmetry was destroyed. He stated:

> It is just here that we meet the most serious doubts concerning the replacement of the actual many-body problem by a one-body problem with a central field for each electron, even as a first approximation.

He then went on to say, that in order to make some headway, a simplifying assumption was necessary. He proposed that the potential for a core electron be the "total potential of the nucleus and of the whole electronic distribution of charge, less the potential from the centrally symmetrical field calculated from the distribution of charge of that electron *averaged over the sphere of each radius.*" The latter was simply $P^2(r)$.

For numerical work, it was convenient to start from an initial field for the core. Then the computational process for determining the core could be expressed diagrammatically as:

Initial Field
⇓
Initial Field corrected for each core electron
⇓
Solution of Wave Equation for each core electron
⇓
Distribution of Charge
⇓
Final Field

If the final field was the same as the initial field, the field was called "self-consistent." This 1927 paper shows that Hartree was thinking not only of equations but also in terms of a process whereby a problem could be solved. In fact, the above is similar in some respects to the flowchart notation used extensively some 20-30 years later when computers were being developed.

In the discussion of practical details, Hartree again introduced physical quantities that assured that a calculation dealt only with slowly varying functions. The spherically symmetric field was expressed as $Z(r)/r^2$ where $Z(r)$ was the "effective nuclear charge at radius r." Similarly, the potential was expressed as $v(r) = Z_p(r)/r$, where Z_p was the "effective nuclear charge for the potential at radius r".

For a many electron atom with nuclear charge N, he derived the expression

$$Z(r) = N - \sum_{nl} q(nl) Z_0(nl, nl; r) \tag{4.3}$$

where the sum was over all the nl electron groups in the core, $q(nl)$ was the number of electrons in the group, and

$$Z_0(nl, nl; r) = \int_o^r P(nl; s)^2 ds \tag{4.4}$$

was the electron charge within the radius r. Then, from these relations, he derived the expression

$$-\frac{dZ_p(r)}{dr} = \frac{Z(r) - Z_p(r)}{r} \tag{4.5}$$

The function $Z(r)$ varied slowly from the nuclear charge N at $r = 0$ to a constant at large r, and could be tabulated at relatively large intervals. Two decimal places were usually kept in his earliest work. The function $Z_p(r)$ could readily be obtained from a numerical solution of the differential equation by inward integration since, for large r, $Z(r) = Z_p(r)$. Note that the entire solution involved only integration and the solution of differential equations, both of which Hartree could perform readily using numerical methods.

Hartree presented results for the ground state of helium, obtaining $\epsilon = 1.835$ atomic units (or Hartrees) whereas the observed value was 1.81 in the same units. What really interested him though, was the case of rubidium where the core consisted of the $1s^2 2s^2 2p^6 3s^2 3p^6 3d^{10}$ group of closed shells. The case of $3d$ with $l = 2$ intrigued him in that the function $v(r) - l(l+1)/r^2$ could conceivably have *two* maxima. He analyzed this case and related it to penetrating and non-penetrating orbits. A repeat of this calculation, using a computer program, shows that the self-consistent 3d orbital had associated with it a function $v(r) - l(l+1)/r^2$ with only one maximum. However, during the SCF process, it may well happen that potentials with two maxima arise, resulting in radial functions with additional nodes, possibly leading to convergence problems. Finally, he presented the results for the sodium ion (Na$^+$) and the chlorine anion (Cl$^-$) that he was using in a paper with I Waller and R W James.

Hartree believed that, in some sense, his wave functions for many-electron systems were "best." It was left to John Arthur Gaunt [1], a student of Fowler's, to study the accuracy of Hartree's atomic fields. Gaunt used perturbation theory to first order in the wave functions and to second order in the energies. He performed calculations first on helium and then the general atom. His conclusion was that, of all methods of approximation to the wave function and energy of a complicated atom, Hartree's method was the most generally effective in reducing the first-order error of the wave function and the second-order error in the energy. He read his paper to the Cambridge Philosophical Society on February 27, 1928.

Another paper that followed quickly was a paper by J Hargreaves [2], read on December 10, 1928 also to the Cambridge Philosophical Society. This paper was on the dispersion electrons of lithium, computing a quantity denoted by f, a symbol now common in spectroscopy and referred to as the oscillator strength. He used some of Hartree's calculations to investigate the f-values for the principal series in lithium which meant computing both bound- and free-electron wave functions. In this way he was able to verify the f-sum rule for Hartree's wave functions, namely $\sum f = 1$. Similar calculations had been done by Bjorn Trumpy [3], who worked at the Technical University at Trondheim. Trumpy used a field obtained in a different way. His results for members of the $2s \to np$, $n = 2, 3, \ldots$ series were similar, but the results for $2s \to 3p$ differed by a factor of 6. This illustrated very clearly that oscillator strengths could be very sensitive to the wave function, even in simple cases.

Hargreaves also was Fowler's student. A lot of new ideas were being explored by students associated with Fowler, who were stimulated by Hartree's practical approach to the new quantum mechanics.

Collaborating on X-ray reflection

Of course, Hartree was not the only one who needed to reformulate his theory. Ivar Waller (1898 - 1991) obtained his DSc in 1925 from the University of Uppsala and then assumed a position there as Lecturer. Using Bohr's theory, he had developed a theory of X-ray scattering and the effect of temperature on the intensity of the reflection of X-rays by crystals [5].

In 1919 after WW I, Reginald William James (1891 - 1964) and William Lawrence Bragg (1890 - 1971) both joined Manchester University but in different positions. Bragg had been awarded the Nobel Prize in 1915 along with his father, William Henry Bragg "for their services in the analysis of crystal structure by means of X-rays," and was appointed as Langworthy Professor of Physics, whereas James was appointed Lecturer. There, under the guidance of Bragg, James worked on X-rays and crystal physics.

Early in 1927 (or possibly earlier) Waller and James collaborated on temperature effects on X-ray reflection for sodium and chlorine in rock-salt crystals. Of interest were the "F-curves" (atomic scattering factors) from theory and experiment. Though the results were acceptable and a paper published [6], they decided to ask Hartree to join them in an attempt to collate the results of four separate lines of research:

(1) the formulation (by Waller) of a theory for scattering of radiation by an atom on the basis of quantum mechanics
(2) the calculation (by Hartree) of the Schrödinger distribution of charge in the atoms chlorine and sodium
(3) the measurement of scattering power of sodium and chlorine atoms in rock-salt crystals for X-rays at a series of temperatures (by James)
(4) the theoretical discussion of the temperature factor (by Waller).

Hartree was already thinking about methods for solving Schrödinger's equation, but as a new PhD he must have been delighted to have his expertise recognized and, at the same time, be able to have his emerging ideas tested in an application. On November 15, 1927 he wrote to Waller:

> I have been glad to collaborate with you and James in this paper, altogether it strikes me as a very satisfactory piece of work: the final agreement between observation and experiment is almost too good!

The paper was submitted on December 22, 1927 by Bragg to the Royal Society, only weeks after Hartree presented his papers to the Cambridge Philosophical Society. We will see later that this collaboration established a "Manchester connection" for Hartree.

Hartree maintained close contact with Waller, inviting him to Cambridge. In a letter of March 1, 1928 he wrote:

> Now that I have developed a method for calculating atomic fields and distribution of charge, various people have been asking me to calculate results for the various atoms in which they are particularly interested; and as I think I can probably do the work more quickly than they ..., it seems best I should. So I have been busy mostly with pure numerical work. One really wants a staff of computers for numerical work in connection with atoms like the astronomers have for doing their numerical work!

Then he closes with the remark that he would like to go to Copenhagen in the autumn for a term, taking his whole family. He was applying for a grant from the International Education Board, but thought that he might go even if it was not granted. A few weeks later, Hartree wrote to Bohr and got the reply that it would "be a pleasure to have you here in Copenhagen."

Hartree immediately applied to the International Education Board for a grant. There was a small problem in that he requested a grant for only three months. During the academic year, 1928-9, he had been appointed University Demonstrator in Physics. He thought he could make arrangements for an absence of three months, but not longer. Finally, Hartree pointed out that there had been a previous exception made in the case of Fowler. The fellowship was approved.

That summer Hartree worked intensely on extending his theories to the new quantum mechanics, trying to understand the implication of Pauli's exclusion principle which required that no two electrons with the same spin could have the same quantum numbers and, at the same time, that the electrons were to be indistinguishable. He knew of the ideas of Heisenberg, who in 1926-7 had shown that for two-electron systems, in the case of opposite spin the space function was symmetric as in

$$\Psi = \psi_1(x_1)\psi_2(x_2) + \psi_1(x_2)\psi_2(x_1) \tag{4.6}$$

whereas for the same spin, it was antisymmetric with respect to the interchange of electrons,

$$\Psi = \psi_1(x_1)\psi_2(x_2) - \psi_1(x_2)\psi_2(x_1). \quad (4.7)$$

The latter could be expressed in the form of the determinant,

$$\Psi = \begin{vmatrix} \psi_1(x_1) & \psi_1(x_2) \\ \psi_2(x_1) & \psi_2(x_2) \end{vmatrix}. \quad (4.8)$$

In these equations, $\psi_i(x_j)$ is the ith one-electron wave function for the jth electron. The problem was how to extend these ideas to more electrons?

Hartree was profoundly stimulated by the work of Walter Heitler and Fritz London [7], who applied quantum mechanics to the study of the hydrogen molecule, H_2. Though there were two nuclei in the molecule, there also were only two electrons. In the case of anti-parallel spins, the wave function had the symmetry properties required by the Pauli principle and the energy of the atom was *lower* than the energy of two hydrogen atoms representing binding or, in the language of chemistry, a bond. But there was another solution that described two electrons with parallel spin, was symmetrical and represented repulsion. Hartree was never interested in just the simplest system and immediately tried to apply these ideas to a more complex case. On July 11, 1928 he wrote to London who was then in Berlin, requesting reprints of his work and expressing admiration. He mentioned that he had considered the radical $(CO_3)^{-2}$, which he had treated as $C^{+4}(O^{-2})_3$ but on reading London's paper, it occurred to him that it could also be $C^+(O^-)_3$. He wondered how one could distinguish between these structures, but also did not see how to deal with a group of four atoms.

They started what London in later years referred to as a "lively discussion." In Hartree's rudimentary German, they discussed the three-electron system for the molecule HeH and also complex atoms where d-electrons might be present. At the end, Hartree apologized for his poor writing in German.

London's reply is not available but he must have explained matters and also suggested that Hartree write to him in English. The next letter, written when the Hartrees were already in Copenhagen, explained that he had become interested in the repulsion between rare gas atoms for which experimental values for deviations from perfect gas laws were available for comparison. He had worked this time on a simpler case of repulsion between two helium atoms, and intended to use the wave functions he had obtained from his "self-consistent-field" method. He hoped that the integrals would reduce to products of double integrals which would not be too tedious to evaluate but had not been successful. Hearing that Egil Hylleraas in Oslo was doing very similar work, Hartree intended to write to him and was waiting for a reply before proceeding further. From this correspondence it is clear that the concepts of spin, the Pauli exclusion principle, and the symmetry properties of wave functions were still not well understood, at least by Hartree. This is also the only indication that Hartree considered problems in the realm of quantum chemistry, problems much easier to investigate with analytic functions than with his numerical ones.

On August 23, 1928 he informed Waller that the family would be arriving in Copenhagen four days later. He went to visit Waller right away and, on September 11, 1928 wrote a a letter describing his return journey from Stockholm via Göteburg. In fact, Hartree spent the term collaborating closely with Waller on X-ray scattering. They investigated the intensity of total scattering of X-rays by atoms distributed at random as in a monatomic gas. They began by considering scattering by helium atoms. This was the first calculation where Hartree used Heisenberg's rule [8], not in the calculation of the field, but in the determination of the atomic scattering factor, the F-curve. Another case considered was argon, consisting of filled shells, each of which was spherically symmetric. Comparison of theory with experiment led them to conclude that experiments at longer wavelengths were desired for which the uncertain relativistic effects would be smaller. By the end of November they were ready to prepare the manuscript.

On November 28, 1928, Hartree informed Waller:

> We expect to leave December 13 ... as the annual Dinner of the Research workers at the Cavendish Laboratory is the 15th, and I should like to get to see the Professor and others before they separate for vacation, and it will also be a good deal more convenient for my wife to have a few days longer to get straight before Christmas. I hope this will not mean missing you.

Of course it does, since Waller cannot leave Uppsala until term is over. Hearing that Waller is free on the weekend December 8-9, Hartree makes a special trip to see him, taking the Thursday night train in order to be well rested by Saturday. Hartree had been writing the paper and a point of particular concern for him was the wave function for an arbitrary number of electrons. He said:

> I have been thinking more about the expression of the wave function as

a product as two determinants as I wrote it, and think I am clear about it now.

He looked forward to reviewing the details with Waller. His ideas must have been accepted, since the published paper contains a section *On the formation of approximate wave functions of required symmetry properties for an atom containing several electrons*, based on the ideas of Heisenberg, Dirac, Wigner, Hund, and Delbrück. Spin was still not introduced directly into the wave function, but there was one determinant for electrons of each spin, that assured the exclusion principle was satisfied. Though the paper was typed before it was time for the Hartrees to leave Denmark, he did not have time to put in all the formulas that needed to be done by hand in those days.

By December 26, Hartree was already writing to Waller describing his return from Stockholm:

> It became very sunny by the time the train got to Norköping, and the frost covered trees looked lovely in the sun. I hoped to do some work on the train, but found it a good deal too hot for serious work.

One might have expected him to say "but the scenery was too beautiful for work."

In his correspondence with Waller, Hartree makes no mention of the Institute, other than to say that towards the end of October, Bohr had invited him to give an account of his work on total scattering. According to Gamow, Hartree was well known at the Institute for his dislike of being warm, that the room he worked in was known as the "Hartree'sche Nordpol" (Hartree's North Pole). Nevill Mott from Cambridge was also at the Institute that term and though he writes extensively about his experiences at Copenhagen in his autobiography [2], Hartree is mentioned only in passing. One day Bohr had taken Gamow, Hartree, and Mott by taxi to the museum, *Prinsens Palae*, to see the "Stone and Bronze Age Collection" that was on exhibit. According to Mott, Bohr was a terrific guide. The museum visit left an impression on Hartree too. In fact, in writing to Bohr (December 21, 1928), he referred to it as the "most outstanding memory of our visit to Copenhagen." For Bohr's hospitality, Elaine and Douglas Hartree sent Bohr a copy of the book, *Downland Man*, by H J Massingham [9]. The book is concerned with the stone age culture in England, but Hartree thought it was more than of local interest.

Oskar Klein had helped the Hartrees find housing and was given the book, *Alice in Wonderland*. Hartree explained (December 20, 1928):

> The writer was a mathematician of about the middle of the last (19'th) century – perhaps rather later – and it is told of him that Queen Victoria was so delighted with "Alice" that she asked him to send her a copy of anything else he wrote and was considerably surprised to receive soon a mathematical text-book!
>
> I'm very fond of "Alice." I've read her many times and always enjoy returning to her again, and somewhere or other, in her adventures, one can find and appreciate comments on almost any subject.

Upon the Hartrees' return, they had exchanged houses with his parents. He went on to say "It feels curious coming back to this rather large house after the compact flat in Copenhagen; I feel I want to apologize for the size of it."

The Gamow who had been visiting the Bohr Institute, was George Gamow, then a student at Leningrad University, working on an hypothesis of a nuclear model. He must have told Hartree that he was interested in visiting Cambridge, because, on the same evening that the family returned home, Hartree had talked to Fowler. When he saw Rutherford a few days later, he was informed that an invitation to Gamow was on its way. This must have been a short visit since, in 1929-30 Gamow was a Rockefeller Fellow at Cambridge University. His views of the nucleus ultimately lead to the theory of nuclear fission and fusion. They must have interested Rutherford a great deal. But this also is an example of Hartree's willingness to become involved in helping promote the interests of others.

Upon Hartree's return to Cambridge, the scattering manuscript was given to Fowler with a request that he consider the paper for submission to the Proceedings of the Royal Society. By January 8, 1929 Hartree had received three pages of comments. In communicating to Waller about the comments Hartree showed a great deal of maturity:

> I think Fowler has not followed closely the work on scattering, and I think some of the points he does not find clear are due to misunderstanding, but I think his comments are valuable just for that reason, as he has read it carefully, so that what he finds difficult, others not specialists in this branch may also find difficult.

All matters were dealt with and the paper was received by the Royal Society on January 19, 1929.

In 1928, Hartree was elected by Christ's College to a Fellowship. This required him to lecture which kept him rather busy. But on March 7, 1929 he had time to share good news with Waller:

> I expect you will be interested to hear that Manchester University has appointed me to the Professorship which Milne left vacant when he went to Oxford. It was a great surprise I was asked, and we are very excited about it. I start work there in October. As soon as this term ends we will be going to look for a house; unfortunately the one the Milnes had just got too small for our family.

The family at this point consisted of a girl, Margaret, and a son, Oliver.

The Hartrees always were very hospitable. If a room was available, visitors were accommodated in their home. During May 1929 R de L Kronig from the Rijks-Universiteit te Groningen, The Netherlands, stayed with them while he was giving a short course on Band Spectra. Several years later, Hartree, de L Kronig and Petersen published a paper on *A theoretical calculation of the fine structure for the K-absorption band of Ge in $GeCl_4$* for which Hartree had supplied the wave functions for chlorine. At first sight, the fact that Hartree is first author is surprising. But closer inspection shows that this application required Hartree to compute, not only bound wave functions, but also wave functions for electrons with positive energies, where the asymptotic phase was important. In the paper he states that:

> The calculation of these phases is often carried out by the approximate method of Jeffreys [10], which was developed independently with special reference to wave mechanical applications by Wentzel [11], Kramers [12], and Brillouin [13], by whose names it is commonly known in such applications.

He did not use that method because he thought it would not be good enough for the range of energies and angular momenta required for the application. He then proceeded to develop an appropriate method, using the "phase-amplitude" method for large r, and also considered the case where $v(r) = 0$. Thus a lot of numerical analysis was needed for this application, justifying first authorship. The phase-amplitude method is still often used for continuum functions in the region of large r. It would appear that Hartree developed the method on his own, but Milne [14] derived the phase-amplitude relationship in 1930 for the purpose of finding discrete eigenvalues.

Once term was finished, Hartree could look forward to attending a conference organized by Pauli in Zurich. He had been invited to give an account of his work on the approximate solution of the many-electron problem. As an example, he would use his work with Waller on the atomic scattering factor. He was delighted to hear that Waller too would be attending, presenting a paper on scattering theory, and assured him that his own talk would in no way overlap with Waller's.

Lindsay and his self-consistency ideas

From a description of these activities, we see that Hartree used the self-consistent-field method to advantage. Why did Lindsay not revise his self-consistency ideas?

Lindsay, in his autobiography (see Chapter 3, [11]), admitted that he did not see how this new point of view could be applied to polyelectronic atoms, and here he claimed he missed a golden opportunity, for "this was just waiting to fall into someone's lap." He continued:

> Why did I not seize the opportunity independently? The answer is a relatively simple one. I did not know enough about the manipulation of partial differential equations to see how the Schrödinger wave equation could be solved by numerical means for a potential energy function represented only by a table of numerical values... Another obvious trouble was that it took me a long time really to understand how the new Schrödinger point of view worked, since I had never had the eigenvalue problem pounded into my mathematics course.

Thus we see how Hartree's background in differential equations played a critical role.

Notes and references

[1] Gaunt J A 1928 *Proc. Camb. Phil. Soc.* **24** 328
[2] Hargreaves J 1929 *Proc. Camb. Phil. Soc.* **25** 75
[3] Trumpy B 1928 *Zeit. f. Phys.* **50** 228
[4] James R W and Brindley G W 1931 *Phil. Mag. Ser.* 7 **12** 81
[5] Waller I 1923 *Zeit. f. Phys.* **17** 398
[6] Waller I and James R W 1927 *Proc. Roy. Soc. A* **A117** 214
[7] Heitler W and London F 1927 *Zeit. f. Phys.* **44** 455
[8] Heisenberg W 1926 *Zeit. f. Phys.* **39** 499
[9] Massingham H J 1926 *Downland Man* (London: J Cape)
[10] Jeffreys H 1924 *Proc. Lond. Math. Soc.* **23** 420
[11] Wentzel G 1926 *Zeit. f. Phys.* **38** 518
[12] Kramers H A 1926 *Zeit. f. Phys.* **39** 828
[13] Brillouin L 1926 *Compte Rendue* **183** 24
[14] Milne W E 1930 *Phys. Rev.* **35** 863

Chapter 5

Advances in atomic theory

Several advances followed rapidly after Hartree's 1927 announcement of his self-consistent field (SCF) theory, whose derivation was based on the intuitive notion of an atom in which electrons moved in a central field.

John C Slater (1900-1976) graduated with a PhD from Harvard in 1923 and immediately received a Sheldon Traveling Fellowship for a year during which he intended to visit Bohr in Copenhagen. Since Bohr was to be in the United States in the Fall of 1923, Slater decided to spend the Fall term at Cambridge, England. He attended some graduate courses, one given by Rutherford, and seminars at the Cavendish Laboratory. He claimed the atmosphere at Cambridge was not that different from Harvard, in that both were stimulating. He admitted not getting to know students since he did not live in College. Apparently Dirac, in later years, reminded him that they had been in the same class, but Slater did not recall that.

Immediately after Hartree's papers describing the self-consistent field appeared, Slater (now on the faculty at Harvard) made a study of their accuracy. He stated, "The first step in discussing Hartree's method is to formulate his process mathematically" which he proceeded to do. He assumed the many-electron wave function was a product of one electron functions and studied how well this product satisfied Schrödinger's equation. The analysis showed that the Hartree equations (and energies) should be corrected slightly, because the distributions were not always spherically symmetric. He went on to show that the estimated size of the corrections was of the order of the errors present in the numerical cases he worked [1]. Thus by the end of 1928 Hartree's method was placed on a firm mathematical foundation. Just before Hartree's departure for Denmark, Slater had sent him a draft manuscript of this paper. Hartree responded with some comments:

> I find the difference of notation makes it difficult to compare exactly what you have done with what Gaunt has done. I find your treatment the cleverer and more attractive; you make it clear what you are doing and what the various quantities involved are. Gaunt's treatment looks analytically heavier...

Unfortunately, we do not know what was in Slater's letter, but can only surmise from Hartree's response:

> Certainly I do not feel at all that you are treading on my toes in working on this subject; on the contrary I am very glad the further this has attracted you and that you and Gaunt have been able to justify the procedure I adopted empirically.

Hartree's many-electron wave function had associated with it the notion of a wave function that was a product of one-electron functions. In 1926-27 Heisenberg had studied the symmetry properties of solutions of the two-electron problem (see Equations (4.6) and (4.7) of the previous chapter). Following Heisenberg's work, several others – Wigner, Hund, Heitler, Weyl – attempted to extend this work to many electron systems using group theory. Indeed, this was exactly the problem Waller and Hartree grappled with in their study of scattering. The application of group theory to quantum mechanics was referred to as the "Gruppenpest" (plague of group theory) by disgruntled individuals who were not familiar with group theory. Slater [2] described his feeling as one of outrage at the turn the subject had taken. He decided to replace the orbitals (one-electron wave functions) represented by three space quantum numbers, by "spin-orbitals" as he called them, that were represented by three space quantum numbers and a fourth quantum number representing spin. Then a determinantal function in terms of spin-orbitals would always be antisymmetric in the exchange of two electrons as well as obey Pauli's exclusion principle.

In a subsequent, very significant paper [3], entitled *The theory of complex spectra*, Slater showed how atomic multiplets (a group of terms according to Slater) [4] could be treated by wave mechanics, without using group theory. As Slater said:

> ...the objects of the present paper resemble closely those aimed at by Heisenberg, Wigner, Hund, Heitler, Weyl, and others who employ the methods of group theory. That method is not used at all in the present calculation, and, in contrast, no mathematics but the simplest is required, until one actually comes to the computation of the integrals.

He then described how wave functions could be expressed as linear combinations of determinants (to guarantee proper symmetry) and one-electron spin-orbitals that depended only on the four quantum numbers, n, l, m_l, m_s. Today these determinants are referred to as "Slater determinants." The expansion coefficients were determined from eigenvalues of an energy interaction matrix. The evaluation of these was greatly simplified if the one-electron orbitals were orthogonal. Philip Morse, who wrote Slater's biography, regarded this as Slater's greatest work [5].

All this appealed to Hartree (letter to Slater, December 18, 1929, from Manchester):

> I'm very pleased at your justification of my self-consistent field method, and especially that you have convinced the "pure" theoretical physicists (I mean the ones like Wigner who have the attitude of a pure mathematician rather than a physicist) that there's something in it – more, I admit, than there was intended to be when I started.

There then followed a discussion of the advantages of analytical versus numerical methods. Hartree believed that the wave function could be wildly wrong at $r = 0$ with an analytic basis. However, with the selection of a proper basis this need not be the case and to this day there are "numerical" methods and "analytical" methods that often use what are called "Slater orbitals."

In the summer of 1929, Slater and his wife, left for a trip to Europe having obtained a Guggenheim Fellowship for six months. First they went to Pauli's conference in Zurich where they met Hartree for the first time. The remaining time was spent in Leipzig with Heisenberg and Hund, who proved to be very interested in Slater's work that, in many ways, was an extension of their work. In a letter to the Editor of *Physical Review*, written in Leipzig, Germany on December 19, 1929, Slater reported that the Hartree equations were closely related to the variational method [6]. If a solution of the form $\Psi = \psi_1(x_1)\psi_2(x_2)\ldots\psi_n(x_n)$ was assumed, the stationary condition of the energy would lead to equations very like the Hartree equations, except that Hartree defined the field in terms of a spherically averaged field. He then mentioned that, instead of one product, the wave function could be assumed to be a linear combination of products so as to have the proper symmetry relations, but did not actually derive any equations.

The Slaters planned to visit the Hartrees in Manchester in January, 1930 on their way to Liverpool, before returning to Harvard. As always,

Hartree was interested in making sure his visitors would have a pleasant time (January 7, 1930):

> I do not know if you and Mrs Slater are fond of music, but if you are, there is an attractive concert on Thursday evening, one of a series given by the Manchester Orchestra which many people consider the best in England. I enclose a programme; would you let me know if either or both of you would like to go?

During this visit Hartree and Slater discussed an idea for improving the calculation of atomic fields by taking account of "austausch" (exchange). Hartree had been wanting to this do for a long time, but had not seen his way through all obstacles. He had worked it out for a three-electron atom, then the academic term began and he had not been able to do more. He became quite discouraged when Fock's paper appeared.

At essentially the same time as Slater was writing his letter to the Editor of *Physical Review*, Vladimir Aleksandrovich Fock (1898 - 1974) presented a paper [7] at a meeting of the Russian Physical-Chemistry Society on December 17, 1929. In his paper, Fock developed a clear, mathematical foundation for the application of the variational principle, first for two-electron systems and then for the many-electron system. In each case he started with a wave function that was the product of one-electron wave functions and showed (as did Slater) that these led to the Hartree method though with a small difference. For two-electron systems an approximate wave function needed to be a sum of products as given by Equation (4.6) or Equation (4.7). In both of these cases, the variational equations included terms not present in the Hartree equations. These terms arise from the second product in the definition of the wave function where the assignment of electrons to orbitals (one-electron wave function) had been interchanged. For this reason, the equations were called "equations with exchange." For the many-electron case, the simple product wave function again led to the Hartree equations. In applying the variational method to wave functions with the proper symmetry Fock proceeded with care. Taking lithium with its $1s^2$ core and one outer electron as an example, he showed that there were several sums of three-electron products satisfying the Pauli principle, but for variational methods, the product of two determinants as proposed by Waller and Hartree was sufficient. He then proceeded to make the assumption that the many-electron wave function was a product of two determinants, one for each spin orientation. He started initially without assuming any orthonormality, but ended by showing that the final equations could be derived from an "energy integral" (more properly, an "energy functional") together with orthonormality conditions of one-electron wave functions. These conditions required the introduction of Lagrange multipliers in the application of the variational principle.

In the notation of the book *The calculation of atomic structures* that Hartree published in 1957, the equations had the form

$$\left[\frac{d^2}{dr^2} + \frac{2}{r}Y(nl;r) - \varepsilon_{nl,nl} - \frac{l(l+1)}{r^2}\right]P(nl;r) =$$
$$X(nl;r) + \sum_{n \neq n'} \varepsilon_{nl,n'l}P(n'l;r) \quad (5.1)$$

together with the orthonormality conditions

$$\int_0^\infty P(nl;r)P(n'l;r)dr = 1 \quad \text{when} \quad n = n'$$
$$0 \quad \text{otherwise.} \quad (5.2)$$

The function $Y(r)$ was defined as

$$Y(r) = N - \sum_{n'l} q(nl)Y^0(n'l,n'l;r) + \sum_k \alpha_{nl,k}Y^k(nl,nl;r) \quad (5.3)$$

$$X(r) = -(2/r)\sum_{n'l',k}{}' \beta_{nl,n'l',k}Y^k(nl,n'l';r)P(n'l';r) \quad (5.4)$$

where $\sum'$ denoted the omission of $n'l' = nl$ and

$$Y^k(nl,n'l';r) = r\int_0^\infty P(nl;s)\frac{r_<^k}{r_>^{k+1}}P(n'l';s)ds. \quad (5.5)$$

In the latter, $r_<$ and $r_>$ are the lesser or greater of r and s, respectively. The coefficients $\alpha_{nl,k}$ and $\beta_{nl,n'l',k}$ depend on the atomic configuration and term energy expression. Though not used by Fock, these were the coefficients that could be derived using Slater's theory for atomic multiplets.

Because of technical difficulties presented by these equations, Fock in collaboration with M J Petrashen (who did the numerical work) were slow in reporting results. The calculations needed to include not only exchange, but also orthogonality of the one-electron radial functions. As their first example, they chose Na with the electrons assigned to $1s^2 2s^2 2p^6 3s$ – not exactly the simplest case. In a study of the numerical solution [8] for this system, they proposed that the wave functions for the core ($1s^2 2s^2 2p^6$) be obtained from the ion, and that the outer electron then be obtained from a linear operator, operating on the radial function for the outer function. This view had the advantage that the integro-differential equation for the outer electron was of eigenvalue type. Initial estimates for the core radial functions were obtained by applying variational methods to analytic functions, with certain free parameters. Green's functions were constructed for the solution of differential equations. Radial functions were tabulated. Checking the one-electron energy parameters, $\epsilon_{nl,nl}$, with results of a present day computer program, agreement was found to be surprisingly good except for the 2s and 2p electrons.

Year	1s	2s	2p	3s	3p	4s	4p
1934	40.6	3.00	1.83	0.183	0.1094	0.0703	0.0501
2000	40.76	3.07	1.80	0.1818	0.1094	0.0701	0.0503

The method, however, was extremely laborious and results were not published until 1934. To explore the effect of exchange, they then selected some simpler cases. Calculations for beryllium-like systems ($1s^2 2s^2$) were performed using analytic functions [9] but the total energy was not as low as the numerical results obtained later by Hartree and Hartree in 1935. Calculations for lithium states ($1s^2 nl$) were done, again assuming the core was unaffected by the outer electron [10]. Oscillator strengths (parameters which describe the absorption and emission processes) were reported and compared with those obtained by Hargreaves neglecting exchange. It is interesting to compare their values with essentially exact values that can be obtained today. Consider the f-values for transitions from $2s \rightarrow np$.

n	without exchange	with exchange	present
2	0.7000	0.768	0.7470
3	0.0104	0.0037	0.00471
4	0.0147	0.0035	
5	0.0051	0.0015	

The significant improvement resulting from the inclusion of exchange is seen, but also evident is the sensitivity of the $2s \rightarrow 3p$ transition where there is a lot of cancellation in the calculation of the oscillator strength.

In the Fall of 1930, Slater was appointed the new Chairman of the Physics Department at the Massachusetts Institute of Technology (MIT) and was planning to develop the Department along the lines of modern physics. "It really looks as though we should start a systematic program of investigating wave functions here," he stated in a letter to Hartree, on November 25, 1931. Their plan was to use orbitals of the form $r^m e^{-a_m r}$,

$m = l+1, \ldots, n$, as an analytic basis, now called "Slater orbitals." He asked Hartree if he had a numerical scheme for equations with exchange, including orthogonality, which greatly simplified the calculation of energy matrices. He had no desire to duplicate what Hartree had done, but at the same time, Hartree's results were not generally available, particularly wave functions. Often only the field $Z(r)$ was published.

Hartree responded (December 21, 1931) saying he was recovering from an illness which had kept him in bed for most of the term – nothing serious but lengthy. Of course, he would be delighted to give Slater all the information about wave functions that he had. Most were in unpublished manuscripts, sent out upon request from various individuals. He had considered publishing the material but was no longer satisfied with the accuracy. He then listed the atoms for which he had calculations (still without exchange): He, Li^+, Be^{++}, O^{+3}, O^{+2}, O^+, O, F^-, Ne, Na^+, Al^{+3}, Al^+, Si^{+4}, Cl^-, Ar (sketchy), K^+, Ca^{++}, Cu^+, Rb^+, and Cs^+ (not completed). He mentioned that he had someone (probably Brindley) starting Ti^{+4} and someone at Cambridge undertaking Hg (probably his father), a system that was of special interest to him. He emphasized that not all the work had been done by himself and expressed concern for accuracy.

> I think it is a question not so much of making mistakes as not finding them when they are made; I find it difficult to get some people to take adequate care in checking work.

Hartree remarked that he had thought a fair amount at one time or another about trying to solve the Fock equations, concluding:

> The solution of SCF equations is probably going to be work for babes compared with the solution of the Fock equations.

He had some ideas but agreed with Slater that, at the moment, analytic methods with variable parameters might be best.

The complexity of the "Fock equations" presented too large a challenge for a numerical solution at the time. Instead, papers began to appear that were based on Slater's theory showing how expressions could be obtained for energies of multiplets [3], a set of close lying terms associated with the same nl assignment of electrons. At that point, Hartree's self-consistent field method (without exchange) was used to obtain radial functions, but energies were computed using Slater's ideas. One of the first of such studies was reported by J McDougall of Cambridge on the calculation of terms of the optical spectrum of Si^{+3} for which the core was $1s^2 2s^2 2p^6$ [11], the paper having been submitted by Fowler. Considered were the series (outer) electrons $3s, 4s, 5s$, $3p, 4p, 5p, 6p$, and $3d, 4d, 5d, 4f, 5g, 6h$. It was now shown that the discrepancy in the quantum defects compared with those derived from observation was reduced by about a factor 7, a very satisfying improvement. The discrepancy had been attributed by Hartree to core polarization when analyzing spectra using the earlier Bohr theory, but was now shown to be due to an exchange effect.

The main difficulty McDougall encountered was the evaluation of two-dimensional integrals, now called "Slater integrals." At this point Hartree was already at Manchester but he advised McDougall by providing him with a tractable and efficient method for computing these integrals. It would appear that in working out the three-electron atom, Hartree had already anticipated these problems. Assume $\alpha, \beta, \gamma, \delta$ are labels for nl quantum numbers not necessarily different. Then

$$R^k(\alpha,\beta;\gamma,\delta) = \int_0^\infty \int_0^\infty P(\alpha;r_1)P(\beta;r_2)\frac{r_<^k}{r_>^{k+1}}P(\gamma;r_1)P(\delta;r_2)dr_1 dr_2 \tag{5.6}$$

where $r_<$ and $r_>$ are the smaller and greater of r_1 and r_2, respectively.

Let

$$Z^k(\alpha,\gamma;r) = r^{-k}\int_0^r P(\alpha;s)P(\gamma;s)s^k ds \tag{5.7}$$

$$Y^k(\alpha,\gamma;r) = Z^k(\alpha,\gamma;r) + r^{k+1}\int_r^\infty P(\alpha;s)P(\gamma;s)s^{-(k+1)} ds \tag{5.8}$$

Then

$$R^k(\alpha,\beta;\gamma,\delta) = \int_0^\infty Y^k(\alpha,\gamma;r)P(\beta;r)P(\delta;r)dr. \tag{5.9}$$

At this point the problem had been reduced to *three* different integrals. For numerical work, Hartree found that it was easier (less work) if the first two equations were expressed as differential equations. Differentiating with respect to r and simplifying, he derived the differential equations:

$$\frac{d}{dr}Z^k(\alpha,\gamma;r) = P(\alpha;r)P(\gamma;r) - \frac{k}{r}Z^k(\alpha,\gamma;r)$$
$$\frac{d}{dr}Y^k(\alpha,\gamma;r) = \frac{1}{r}\left[(k+1)Y^k(\alpha,\gamma;r) - (2k+1)Z^k(\alpha,\gamma;r)\right] \tag{5.10}$$

The first could be solved by outward integration, starting with $Z^k(\alpha,\gamma;0) = 0$ and the second by inward integration with the boundary condition

$Y^k(\alpha,\gamma;r) = Z^k(\alpha,\gamma;r)$ at a large value of r.

At Manchester, Murial M Black, Hartree's MSc student, attempted a somewhat more complex case, namely, the oxygen ground configuration $1s^22s^22p^4$ and its three terms, $^3P,^1D,^1S$, and several ions: O^+ $1s^22s^22p^3$ with terms $^4S,^2D,^2P$; O^{++} $1s^22s^22p^2$ with terms $^3P,^1D,^1S$; and O^{+++} $1s^22s^22p$ ^2P. A paper describing this work, appeared a year later in 1933. Like the earlier McDougall paper, a large part of this work was devoted to evaluating the two dimensional integrals that arose. Another factor that needed attention (already dealt with by McDougall) was orthogonality of the orbitals. In the Hartree approximation, radial functions were not required to be orthnormal, something that Slater's method for computing energies relied on. Once all integrals had been evaluated, ionization energies, and term separation could be compared with observation. Though agreement using Slater's theory for complex spectra was generally good, it was noted that the calculated ratio of $(^1D-^1S)/(^3P-^1D)$ energy was 1.5, whereas the observed ratio was 1.04 in O^{++}. But as Hartree said, "For an approximation which gives an error of this magnitude in the ratio, the general agreement between absolute values of calculated and observed intermultiplet separations was perhaps even surprising." Hartree admitted to Slater that this work brought home forcefully the advantage of dealing with functions of a definite analytical form but, in spite of this, he continued to use numerical functions.

William Hartree and calculations with exchange

In 1933, W Hartree gave up research in physiology, no longer being able to spend long hours on his feet in the laboratory. He turned to his son, suggesting that he could help by doing some computational work for him. Being extremely busy at Manchester, Douglas accepted his father's offer and a most productive collaboration began.

In 1935 Hartree and Hartree reported results for the ground state of beryllium ($1s^22s^2$) including exchange. They pointed out that the practical handling of the "Fock" equations (as Hartree called them) was considerably more complex than the SCF problem without exchange. The paper also mentioned that the procedures given in the paper already had been developed when the Fock and Petrashen paper appeared. The fact that this is mentioned at all suggests that Hartree still had lingering regrets about not having published his results earlier. An important technical point raised in this paper was the arbitrariness of the solutions of Fock's equations. They referred to Koopmans [12] result which showed that the case where the off-diagonal Lagrange multiplier is set to zero was the best, but preferred not to use that scheme in their work.

Together with his father's help, the calculations for Cl^- and Cu^+ were repeated in order to observe the effect of exchange. Negative ions were of interest because of the small values of the binding energy, but Cl^- was of special interest in that there was a considerable difference in the computed and observed diamagnetic susceptibility. It also was the negative ion for which the atomic scattering factor for X-rays had been measured most accurately. They found a considerable contraction of the $3p^6$ group, and, in contrast, a very small contraction of $3s^2$. The improvements in the calculated susceptibility and electric polarizability were considerable. The Cu^+ atom was chosen as the next system to attempt, in that it appeared the exchange effect was likely to be large since in earlier work, results for the 3d electron had been so sensitive to the field. Indeed, the calculations remained troublesome, but as in Cl^-, there was considerable contraction of the outermost (nl) group accompanied by a much smaller contraction of the other shells. The diamagnetic susceptibility was found to be in good agreement with experiment.

On April 2, 1936 Hartree sent Slater a copy of the paper on Fock's equation for Cl^-, mentioned that Cu^+ was almost finished, and that he planned to attempt Fe next. He would also like to repeat the work of oxygen including exchange in the determination of the radial functions. He referred to Leo Goldberg at the Harvard College Observatory who was contemplating the solution of Fock's equations on an extensive scale in order to be able to perform intensity calculations needed in astrophysics. With obvious excitement, he also confided that he had been asked if he wanted to spend a year at Princeton. He saw no possibility of a long visit for two years or so.

The effects of configuration interaction

In 1937 Bertha Swirles was a member of Hartree's staff, teaching Mathematics at the University of Manchester and able to assist with the more theoretical tasks of deriving energy expressions and interaction matrices.

Hartree and Black had noted that the ratios of term separations in O, O^+, and O^{++} differed substantially from observed. Now Hartree and

Swirles decided to investigate the possibility that, for an accurate description, "configuration interaction" was needed, something discussed in Condon and Shortley's *Theory of atomic spectra* [13]. They assumed the many-electron wave function was a linear combination of the $2s^2 2p^q$ and $2p^{q+2}$ configurations, where $q = 4, 3, 2$ for O, O$^+$, and O^{++}, respectively. Using the same SCF radial functions as before, energies for the three terms of each of O, O$^+$, and O^{++} were computed. They found that all ratios, but particularly the $(^1D - {}^1S)/(^3P - {}^1D)$ ratio in O and O^{++}, were over-corrected. They concluded that it would be desirable to include configuration interaction in the equations from which the radial wave functions were obtained.

They proceeded to investigate matters carefully, first the effect of exchange on the radial functions and then the effect of both exchange *and* the two configuration approximation for the wave function. They called the latter method "superposition of configurations." Bertha derived all the necessary coefficients for the radial equations for the different cases. For the solution of the differential equations, they called upon William Hartree for assistance. William Hartree went about solving, first the differential equations with exchange for O, O$^+$, and O^{++}, and then the differential equations with both exchange and superposition of configurations, but limiting the work to O$^+$. The latter was found to have an almost negligible effect on the radial functions to the three-decimal place accuracy of their calculation. The resulting Hartree, Hartree, and Swirles paper (published in 1939) was an important milestone in Hartree's work on atomic structure calculations.

The results, however, were not entirely satisfactory. In O the ratio now agreed more closely with experiment – no doubt this was reassuring – but in O^{++} the ratio was still over-corrected though slightly improved and O$^+$ did not change appreciably. In 1942 Swirles (now Bertha Jeffreys) published a note [14] on an anomaly in the spectrum of O^{++}. She investigated the effect of allowing the $2p$ wave functions to be different for the two configurations and concluded that "an entirely fresh approach is necessary."

In 1963 the present author (then Charlotte Froese) was developing a computer program for performing such calculations and tried to repeat the oxygen work, without success. Fred I Innes of the Spectroscopic Studies Branch, Hanscom Field, Cambridge, MA (USA) too could not reproduce the results for O$^+$. Innes contacted Swirles in 1963 about this matter [15]. At that time she had William Hartree's work books. All his work had been done in pencil for easy erasure, in exercise books of 52 pages ruled in squares. There she found that he had been given a constant of $\sqrt{3}/3$

rather than the correct value of $\sqrt{2}/3$. It was a coefficient in an expression for the interaction between the two configurations. However, since the effect of superposition of configurations on radial functions was essentially negligible, this did not result in a significant error in the final result. Innes pointed out a misprint in the paper for O^{++} and also raised a question about phase conventions. In atomic theory, different phase conventions are encountered and need to be applied consistently within a calculation which Swirles believed she had done. There must have been other errors. The final values I found [16] in 1967 were very similar to the calculations first performed by Hartree and Swirles in 1937 where energy interaction matrices were computed using self-consistent field radial functions. Today we know that many more configurations are needed for an accurate reproduction of the observed ratio of term differences.

Hand calculations of such complexity are, of course, prone to errors, involving the recording of pages and pages of intermediate values. Douglas Hartree constantly was on the lookout for errors that would result in functions that were not smooth, but an error in a constant for the problem would have been impossible to detect by this method. Furthermore, it always is more difficult to be the first to do a calculation because comparison with experiment is not reliable for such approximate theory.

The person who followed up on these ideas in the next few decades was Jucys. Adolfas P Jucys (1904-1974) [17] graduated from Kaunus Vytautas Magnus University, in Kaunus, Lithuania in 1931 where he became interested in quantum mechanics. Desiring contacts with leading scientists in the field, The Council of The Faculty of Sciences sent him to Manchester University for six weeks during the summer of 1938 to work with Hartree. During that time he studied the modern methods of quantum mechanical calculations. Hartree suggested that he investigate the calculation of the Y^k functions as solutions of second-order differential equations. In 1939 Jucys published the paper *Self-consistent field with exchange for carbon* [18], thanking both D R and W Hartree for advice concerning computation. It is a remarkable paper, reporting SCF results for three terms – ^3P, ^1D, and ^1S – including both exchange and configuration interaction. In 1939 he also visited Fowler at Cambridge University. It was during this visit that World War II broke out and Jucys needed all his ingenuity to return to Lithuania. In 1941 he published his PhD dissertation *Theoretical investigation of ions C^{4+} and C^{2+} and of neutral C*. He described there not only the results of the calculation of wave functions and energies for the carbon atom but also a detailed description of the method of solution of the Hartree-Fock equations and of obtaining radial functions.

In 1944, Jucys became the Chairman of the Department of Theoretical Physics at the State University of Vilnius. He spent 1949-51 in the Leningrad Branch of the V Steklov Mathematical Institute of the USSR Academy of Sciences, where he had the opportunity to work with V Fock. In 1951 he submitted his "Doctoral" thesis (beyond the PhD) on *Some refined methods of quantum mechanical calculations of an atom* at Leningrad University. It was a source of ideas for himself and his younger colleagues for several decades. Returning to Vilnius he was promoted to full professor in 1952 and elected to the Lithuanian Academy of Sciences the following year.

To those atomic physicists he met, he was proud to relate that he was the only scientist who had worked with both Hartree and Fock.

Relativistic theory

Dirac was a classmate of Hartree's. Both were undergraduates at St John's College, and then Fellows (Hartree in 1924 and Dirac in 1927). They obtained their degree in the same year, with Dirac a student of Fowler's. In other respects, these two scientists differed markedly.

Paul Adrien Maurice Dirac (1902-1984) was born in Bristol in 1902. Thus he was too young to have his education interrupted by World War I. He had obtained a degree in engineering from Bristol University but stayed an additional two years, studying mathematics before transferring to St John's College, Cambridge in 1923. There he was introduced to work on the atom by Bohr and Rutherford. On obtaining his PhD, he began travels to major centers in Europe, visiting leading figures in the developing field of quantum mechanics, such as Bohr, Pauli, Born, and Heisenberg. In 1928, he formulated his relativistic theory of an electron. The importance to atomic theory of his work lies in his famous wave equation, which introduced special relativity into Schrödinger's equation. In 1933, he together with Schrödinger, were awarded the Nobel Prize in physics "for the discovery of new productive forms of atomic theory."

While Hartree and Dirac were undergraduates, the mathematical physicist, Charles Galton Darwin (grandson of Charles Darwin of *Origin of the Species* fame) was a Fellow at Christ's College. There is no indication of any interaction with either Hartree or Dirac at this time, and Darwin left

Cambridge in 1923.

With the new quantum mechanics flourishing, Darwin too turned to atomic physics research, applying ideas about relativity and spin to the motion of the electron. In 1928, he helped many scientists to better understand Dirac's work that was expressed in terms of non-commuting algebras. He wrote an important paper [19], the object of which was to take Dirac's system and treat it by the ordinary methods of wave calculus. It was also of interest to Darwin to see the relationship between Dirac's theory and the equations he had derived himself about a year before Dirac's result appeared. He then went on to show how Dirac's four-rowed matrices imply four radial functions which satisfy certain simultaneous differential equations. This would have pleased Hartree who was much more familiar with differential equations than commuting algebras.

Hartree's name is not usually associated with relativistic theory but, in fact, he was a great admirer of Dirac. While Hartree was at the Bohr Institute in the Fall of 1928, waiting for the paper with Waller to be typed, he thought about the application of his central field method to the study of atoms using Dirac theory. At the core of Hartree's self-consistent field method was the property of spherical symmetry of groups of electrons. In his first paper on Dirac's theory, he studied these symmetry properties of many electron systems using wave functions expressed approximately as products of one-electron functions. In Dirac theory, every electron except an s electron which is already described by a spherically symmetric charge distribution and zero angular momentum, now has two charge distributions – one for each of the two possible spin orientations, generally denoted as either up ($\uparrow$) or down ($\downarrow$). In non-relativistic theory, by Pauli's exclusion principle, there can be only one electron for each spin, with the quantum numbers n, l, and m_l and, since $-l \leq m_l \leq l$, there are $2l+1$ electrons of a given n, l, and a given spin orientation. These are called a "complete half-group," whereas the $2(2l+1)$ electrons of either spin directions form a "complete group." The distribution of charge for a complete-half group is spherically symmetric and therefore also for the complete group. In modern terminology, the latter is referred to as a "filled subshell." The rare gases consist of a series of subshells, all of them filled.

In Schrödinger theory, spin does not enter into the wave equation, only in the antisymmetry requirement. In Dirac theory, the spin is an essential part of the description of a wave function and its equation. For a given l, the quantum numbers are $j = l \pm 1/2$, and $-j \leq m_j \leq j$. Hartree showed (in 1929), that for electrons obeying Dirac's equation, the $2j+1$ electrons with the same n, l and j can be divided into two halves ($m_j < -1/2$ or $m_j > 1/2$), and for each group the distribution of charge is spherically symmetric. In his letter to Waller (November 28, 1928) he considers the groups $2s^2 2p^6$ and lists the spherically symmetric distributions of charge, first in Schrödinger mechanics,

No	(nl)	m_l	m_s
1	$2s$	0	↑
1	$2s$	0	↓
3	$2p$	$0, \pm 1$	↑
3	$2p$	$0, \pm 1$	↓

where "No" is the number of electrons in the group, and then in Dirac mechanics

No	(nl)	j	m_j
1	$2s$	$1/2$	$1/2$
1	$2s$	$1/2$	$-1/2$
1	$2p$	$1/2$	$1/2$
1	$2p$	$1/2$	$-1/2$
2	$2p$	$3/2$	$3/2, 1/2$
2	$2p$	$3/2$	$-3/2, -1/2$

So there are 6 separate spherically symmetric distributions in Dirac mechanics, compared with only 4 in Schrödinger mechanics.

At Manchester University Hartree undertook very few calculations himself, but Bertha Swirles remembered clearly that Hartree made the suggestion for extending the self-consistent field method to the Dirac equation on the platform at Euston Station (London) in the autumn of 1934 when they were returning to Manchester from a conference on the nucleus [15]. She proceeded to derive the equations with exchange by an extension of the variational method proposed by Slater [3]. In those days, angular momentum theory had not been developed and just deriving the expressions for the interaction of charge distributions was a major undertaking. Her paper on this subject was published in 1935 [20] tabulating these quantities for complete sub-groups, including d-electrons. As an example, equations were derived for the $2p$ ($j = 1/2$ and $3/2$) radial functions in neutral copper. In this first paper, only the Coulomb interaction was considered along with exchange. Her second paper [21] extended the theoretical development to include the interaction of electron spin and retardation (Breit interaction). A calculation for the $1s2p$ ^3P term of helium was undertaken using an analytic basis. Though Hartree's name does not appear on any of these papers, it is clear he was closely involved and by today's standards, would have been second author. In his review paper on *The calculation of atomic structures* (1948), his appendix includes a number of corrections to Bertha Swirles' paper on *The relativistic self-consistent field* [20].

In Chapter 7 we will see that Hartree kept in close touch with Lindsay and that several theses resulted from this contact. In fact, the first numerical relativistic self-consistent field calculations (without exchange) were reported in 1940 by Lindsay's student, Arthur O Williams. They were for Cu^+ and the numerical work was done on a "rented computing machine." At that time, Cu^+ was the heaviest atom or ion that had been treated both without and with exchange *non-relativistically*. As a result of William's work the relativistic effect could be compared with the exchange effect.

It was not until quite some time later that Hartree suggested a more ambitious undertaking. In 1954, he assigned the calculation of relativistic radial functions, *without* exchange, to a new PhD student, David F Mayers. The atom was mercury (Hg). It is an atom containing only filled shells, which simplifies matters considerably, but the calculation also required the determination of 22 relativistic wave functions that needed to be iterated to self-consistency. The calculations were to be done on a new electronic, digital, stored program automatic computer, the EDSAC, built at the Mathematical Laboratory in Cambridge under the direction of Maurice V Wilkes.

The calculations were challenging for the EDSAC in that the machine initially had only 1024 short (17 bit) words. In due course, Mayers published his results [22]. As was customary at that time, the radial functions were published as well as the field. A comparison was made of the relativistic and non-relativistic radial densities from the $2p$ group. The former were much more contracted with the charge distribution closer to the nucleus. The total radial charge distribution was also tabulated. An increase of charge near the nucleus was clearly evident. Though it is not clear from the data presented, Mayers states that "even the contribution from the $4f$ group is affected, to a larger extent than expected" by the relativistic theory.

In later years, Mayers wrote a program for relativistic calculations based on Dirac theory and including exchange. This was the basis for the work taken over by Ian P Grant at Oxford. The program, ultimately called GRASP for General Relativistic Atomic Structure Package, was widely distributed [23]. It included an angular momentum package based on theory developed by Jucys and his group.

In spite of Bertha Swirles early work, with Hartree's collaboration, relativistic calculations with exchange are often referred to as "Dirac-Fock" (DF) or the multiconfiguration version as "multiconfiguration Dirac-Fock" (MCDF), though the terms "Dirac-Hartree-Fock" (DHF) and "multiconfiguration Dirac-Hartree-Fock" (MCDHF) are also used.

Notes and references

[1] Slater J C 1928 *Phys. Rev.* **32** 339
[2] Slater J C 1975 *Solid State and Molecular Theory: A Scientific Biography* (New York: Wiley-Interscience) which includes also a good review of the development of atomic theory.
[3] Slater J C 1929 *Phys. Rev.* **34** 1293; *ibid* 1312
[4] Slater's use of the word "multiplet" for the levels of a group of terms of a configuration, is not common in atomic physics today where multiplets are the closely spaced levels of one term.
[5] Morse P M 1982 *Biographical Memoirs* **43** (National Academy Press, Wash. DC) pp 297
[6] Slater J C 1930 *Phys. Rev.* **35** 210
[7] Fock V 1930 *Zeit. f. Phys.* **61** 126
[8] Fock V and Petrashen M J 1934 *Phys. Zeits. Sowjet* **6** 368
[9] Fock V and Petrashen M J 1935 *Phys. Zeits. Sowjet* **8** 359
[10] Fock V and Petrashen M J 1935 *Phys. Zeits. Sowjet* **8** 547
[11] McDougall J 1932 *Proc. Roy. Soc. A* **138** 550
[12] Koopmans T A 1934 *Physica* **1** 105
[13] Condon E U and Shortley G H 1935 *The Theory of Atomic Spectra* (Cambridge: Cambridge University Press)
[14] Jeffreys B 1942 *Proc. Camb. Phil. Soc.* **38** 290
[15] Jeffreys, Bertha Swirles 1987 *Comments At. Mol. Phys.* **20** 189
[16] Froese C 1967 *Proc. Phys. Soc.* **90** 39
[17] Brian R Judd wrote a short obituary that appeared in 1974 *J. Opt. Soc. Am.* **64** 1026-7 whereas Z Rudzikas published an article *Self-consistent field method in various approximations: 'forgotten ideas'* in 2000 *Molecular Physics* **98** 1205.
[18] Jucys A 1939 *Proc. Roy. Soc. A* **173** 59
[19] Darwin C G 1928 *Proc Roy Soc* **A118** 654
[20] Swirles, Bertha 1935 *Proc. Roy. Soc. A* **152** 625
[21] Swirles, Bertha 1936 *Proc. Roy. Soc. A* **157** 680
[22] Mayers D F 1957 *Proc. Roy. Soc. A* **241** 93

[23] Dyall K G, Grant I P, Johnson C T, Parpia F A, and Plummer E P 1989 *Computer Phys. Commun.* **55** 425

Chapter 6

Radio waves in the atmosphere

In 1895, Guglielmo Marconi was the first to successfully transmit and receive radio signals and, by the end of 1901, the first to send a signal across the Atlantic from a transmitting station in Cornwall, England to Newfoundland, Canada, a distance of 2100 miles. He did so to prove that wireless waves were not affected by the earth's curvature. Soon wireless telegraphy developed into a means of communication.

As a teenage student at Bedales, telegraphy was one of Hartree's hobbies. In an article he wrote for the December 1913 issue of the *Bedales Chronicle*, he mentioned that it was not an expensive hobby for someone who could build part of the apparatus. Of course, it was necessary to learn all the Morse codes, which was not difficult, he claimed, though the commercial rate of twenty words a minute was rather fast. The weather report from Paris, for example, was only 4-5 words a minute which he thought was quite easy to follow. He described the sounds heard in a receiver as going from "high and almost screeching noise, to a clear, deep musical note." He then listed the messages sent every day from Paris: the time signal, the weather report, and at 8 o'clock in the evening, the evening news at a rate of about 12 words a minute. He mentioned that there also were plenty of other messages like "Have you accepted the invitation to come to dinner to-morrow?"

Hartree continued to be fascinated by "waves in the ether" while at Cambridge. His first research paper after graduation, published in 1923, was in some sense a mathematical exercise in solving Maxwell's equations analytically for wave propagation under certain conditions. It was a study of the properties of a disturbance at the surface of separation of two transparent media. He had hoped it would have a bearing on the mechanism of the photoelectric effect, but it did not.

The scientist who wanted to completely understand the propagation of radio waves was Edward V Appleton (1892 - 1965). It was generally known that diffraction alone could not explain the propagation of waves around a spherical earth. In 1902 Kennelly and Heaviside independently suggested the existence of a conducting layer in the upper atmosphere that would reflect the waves. Through a series of experiments Appleton intended to confirm the existence of the layer and study its properties.

During World War I, Appleton had joined the Royal Engineers and was posted to the Signals Station, where he was introduced to radio communications [1]. His interest was stimulated. Upon returning to Cambridge, he persuaded Rutherford to allow him to continue research on non-linear vacuum tube oscillations along with more studies of the nature of the atmosphere. At a November 28, 1924 meeting of *The Royal Meteorological Society*, he presented a paper in which the facts for long and short wave transmissions were summarized. They showed that the atmosphere exerted a variable and usually favorable influence on wave propagation. In the paper, he gave a formula for the refractive index of plane waves in the direction of the magnetic field that was extended the following year in collaboration with M A F Barnett, a research student [2], to waves perpendicular to the field. By now, Appleton was Wheatstone Professor of Physics at London University. With the co-operation of the British Broadcasting Corporation, Appleton and Barnett were able to confirm that in sending signals from London to Cambridge, two signals were received. One was direct and the other was one that had been transmitted upward and then reflected back towards the ground. The reflection was from a layer in the upper atmosphere, now called the ionosphere, but then known as the Kennelly-Heaviside layer. Appleton reported on his work at the General Assembly of the International Union of Scientific Radio Telegraphy, held in Washington in October 1927. Only abstracts were published and the equation of the refractive index was given without proof, assuming that absorption was omitted. Thus wave propagation in the atmosphere and the refractive index became of immediate interest.

Hartree had an idea and investigated *The propagation of electromagnetic waves in a stratified medium*. The Heaviside layer (as he called it) formed a refracting medium about whose structure little was known, but he thought that as a first approximation it could be treated as a stratified medium. He attempted to answer the question:

> Given two media with a transition layer in which the refractive index

varies continuously from the value of one medium to that of another, how thin must the transition layer be relative to the wavelength of the waves in order to give appreciable reflection; and for waves incident on a layer of small refractive index on both sides, ..., how thick must the layer be in order that the reflection shall be effectively total?

He went on to say that the equations for the propagation of electromagnetic waves in a stratified refracting medium could be derived, either directly from Maxwell's equations or by considering the electromagnetic disturbance at a point as the sum of the incident wave and the integrated effect of wavelets scattered by an element of volume of the refracting medium. The latter, in effect, was an integral equation method in which Maxwell's equations were integrated to include boundary conditions. He considered the latter of some interest in that the scattering process was separated out and need not be treated classically. In this paper only plane waves were considered. The work was largely analytical, though some calculations, possibly on a slide-rule, were needed. Three cases were considered, and one was deemed by Hartree as being of some importance as a first approximation to the conditions in the Heaviside layer. The paper was received by the Cambridge Philosophical Society on October 8, 1928, just at the time when Hartree was starting his three months stay in Copenhagen.

After his move to Manchester, Hartree continued to follow up on some of Appleton's ideas. In February 1930, he informed Waller that all the serious research he had done since coming to Manchester, had been mainly on radio waves. It is possible that Hartree had been in contact with Appleton because, on September 29, 1930, he wrote to Appleton (the italics are added):

> I have been looking into the discontinuity for propagation along the direction of the magnetic field *which you mentioned*; it's there all right, if you neglect the damping, I can't see any particular reason why it should be there, but then I think in any case one can't see the detail how the refracted wave comes to be what it is.

He also enclosed a typescript of another manuscript he had written on *The propagation of electromagnetic waves in a refracting medium in a magnetic field*. Appleton replied promptly (including a correction to his Washington paper which appeared to have to do with the sign of a factor of (1/2) to which Hartree replied (October 12, 1930):

> You ask for a treatment of the polarization term by retarded potentials; I think that is effectively what I have done, except that I've gone one

step further back.

In this second paper, read to the Cambridge Philosophical Society on December 8, 1930, the effect of the magnetic field was treated in detail and a formula for the refractive index derived using a vector and tensor notation. He claimed the final result agreed with the one given in Appleton's Washington Radio Conference (1927). At one level it does, but buried in the formula is a parameter β which Appleton had left unspecified. Hartree, relying on a Lorentz treatment for the electrons in an ionized medium, argued for $\beta = 1/3$ but Sydney Goldstein, an applied mathematician who at the time had a position at both Cambridge and Manchester, had argued for $\beta = 0$ [3]. The analytic derivation of β depended on physical assumptions and the correct value could only be determined with confidence from observation. Hartree felt the issue should be discussed and showed the difference in predicted behavior for the two values, particularly for long waves.

Thus a long-standing discrepancy appeared and it was clear that the matter was more complex than realized at first. On the theoretical side, Darwin [4] wrote two papers and in both concluded $\beta = 0$. Booker and Berkner [5] made observations that supported $\beta = 1/3$. There were many others. Hartree, understandably upset, on April 4, 1933 admitted to Appleton:

> ...It is, after all, a side line of mine; I took it up originally because I saw a problem that seemed to want doing, and that led to others, but I haven't the knowledge of it that people have who've been engaged in it as their main line of work for years. And at present, as far as I have any time for research, I have enough atomic problems to keep me going for some time, unless someone else gets in at them first.
>
> There is a radio wave problem I want to tackle, as a matter of fact, but it is difficult to be done "by-the-way." That is the question of the Lorentz 1/3. Darwin gave me a shock a little time ago by saying that he wasn't sure after all that it *was* right for free electrons (classically, nothing to do with wave mechanics). He thinks he has shown that for the free electrons in a metal it is not right. ...I think he is right, as far as I follow him, but am not satisfied that his argument applies to the Heaviside layer condition. It is a question that needs *extremely* careful thinking.

Some ideas were developed about "group velocities" and "cooperation" and by August 10, 1936 Hartree was able to say "I see through the whole thing now." The matter was finally resolved in 1941 by an analysis conducted by Newbern Smith (1909-1987) [6] then at the National Bureau of

Standards (NBS) in Washington, DC. At NBS, continuous recordings were available of field-intensity measurements for the period 1935-1940 of signals from a broadcasting station at Mason, Ohio, a distance of 645 kilometers from Washington. By then it had been accepted that there were *two* reflecting layers, called the E- and F-layers. The mean value of β from this study was -0.06 from which he concluded that, at least for that particular data, $\beta = 0$. Thus Hartree initiated a "storm" that lasted for a decade.

Today, the equation he derived for the refractive index is known as the *Appleton-Hartree equation*.

During the period 1928 - 1930, Appleton had investigated various methods of examining the structure of the Heaviside layer. A particular point of importance was the interpretation of the optical path or equivalent path, the subject of another paper by Hartree. This one was submitted to the more prestigious *Proceedings of the Royal Society (London)* (received January 20, 1931) by Appleton since Hartree had not yet been elected a Fellow of the Royal Society. In all his discussions, Appleton had used the "ray treatment" based on rays obeying laws of geometric optics with the idea of interference between rays, a method Hartree considered a good approximation in most cases, but inadequate in a few. For these, it was desirable to have a wave treatment of the phenomena discussed by Appleton and this treatment was reported. At the end of the paper he thanked Appleton for his interest, encouragement, and many helpful discussions. Hartree had made a point and in subsequent papers, such as [7], it was stated that for certain conditions the ray treatment needed to be replaced by the wave treatment, as described by Hartree.

In 1932 Hartree was elected as a Fellow of the Royal Society.

Appleton and Hartree continued to correspond about twice a year until 1937. Figure 6.1 is part of a letter that shows not only Hartree's distinctive writing, but the detail of their correspondence. In the last few letters, Appleton asked him about the photo-electric effect and whether it could be computed using a formula derived for a hydrogen-like atom. Hartree pointed out that to a apply such a formula to a non-hydrogen-like atom required the use of an effective nuclear charge, that the appropriate nuclear charge depended on the phenomena, and that the effective nuclear charge for photo-electric absorption was a rather complicated average he would not presume to guess. But Hartree knew someone at Liverpool who had done some calculations of the photoelectric effect using relativistic wave functions. Calculations with non-relativistic ones would be trivial by comparison. In just over two weeks, Hartree had some results for the photo-

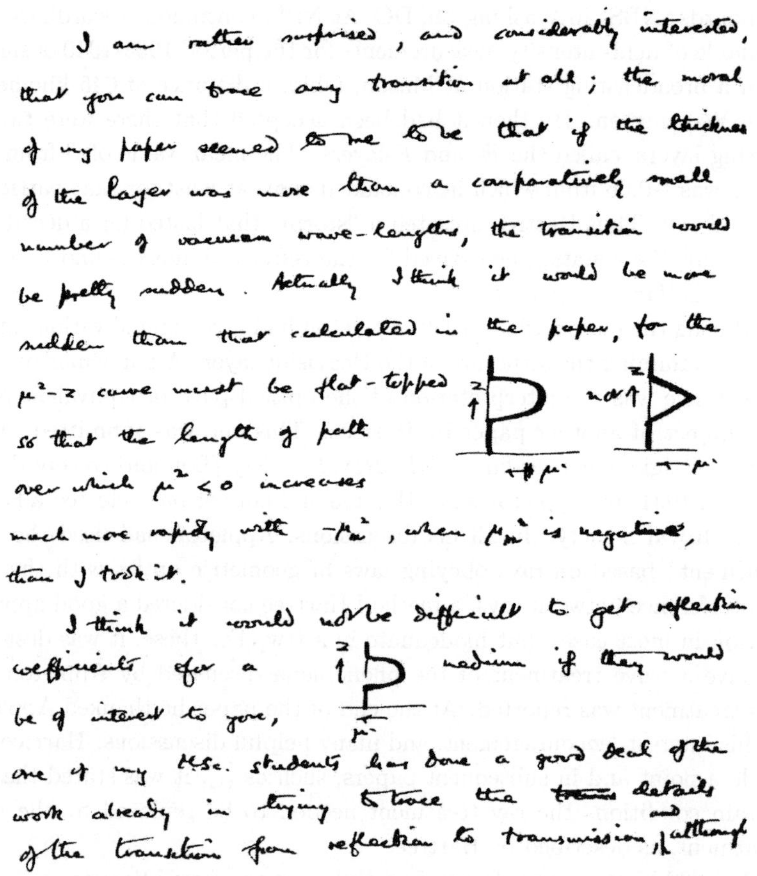

I am rather surprised, and considerably interested, that you can trace any transition at all; the moral of my paper seemed to me to be that if the thickness of the layer was more than a comparatively small number of vacuum wave-lengths, the transition would be pretty sudden. Actually I think it would be more sudden than that calculated in the paper, for the μ^2-z curve must be flat-topped [diagram] not [diagram] so that the length of path over which $\mu^2 < 0$ increases much more rapidly with $-\mu_m^2$, when μ_m^2 is negative, than I took it.

I think it would not be difficult to get reflection coefficients for a [diagram] medium if they would be of interest to you,

One of my M.Sc. students has done a good deal of work already in trying to trace the ~~trans~~ details and of the transition from reflection to transmission, although

Fig. 6.1 Sample of Hartree's writing (Letter to Appleton, dated August 9, 1934) showing technical details.

electric absorption coefficient, computed using hydrogen-like wave functions to share with Appleton. Hartree believed they could be accurate to a factor of 2 or so (April 6, 1937):

> Transition probabilities in the discrete spectrum (except that of the first line in each series) are rather sensitive to the wave functions from which they are calculated, and I think the same would be true of transitions to the continuum; and for things like the $(2p)$ group of oxygen, hydrogen-like wave functions are a pretty poor approximation.

By this time Hartree had proved himself and they seemed to have a good professional relationship as equals. This was to change during the war when Appleton was appointed Secretary of the Department of Scientific and Industrial Research whereas Hartree remained "in the trenches," working on many computational problems.

The wave propagation research prepared Hartree for work during the war on meteorological factors in radio wave propagation. The region of interest was not the ionosphere, but the troposphere or the lowest layer of the atmosphere and the problems investigated were the refraction phenomena for specific variations of the refractive index with height. The work was not analytic but highly computational.

Choices in research directions are customary. Why did Hartree devote all of his free time for research to this problem rather than follow the quantum mechanical path? In a letter to Slater, December 21, 1931 he admits:

> After being forestalled by Fock in my attempt to develop your suggestion of using the variational method to included resonance interactions in the SCF, I felt a desire for a change of subject, and getting into touch at the same time with Appleton of Kings College London, who is interested in reflection of radio waves from the Heaviside layer, I found some interesting work for me there, and am still occupied with one of the problems I've found in the region.

So Fock had pre-empted his research on equations with exchange and Hartree did not want to compete with Fock.

Notes and References

[1] Wilkes M V 1997 *Notes and Records of the Royal Society* **52**(2) 281
[2] Appleton E V and Barnett M A F 1925 *Electrician* **94** 398
[3] Goldstein S 1928 *Proc. Roy. Soc. A* **121** 260
[4] Darwin C G 1934 *Proc. Roy. Soc. A* **182** 152. In a footnote in this paper, Darwin listed several other papers concerned with the problem.
[5] Booker H G and Berkner L V 1938 *Nature* 141 409
[6] Smith N 1941 *J. Res. Nat. Bur. Stand.* **26** 105
[7] Appleton E V and Smith R 1932 *Proc. Roy. Soc. A* **137** 36

Chapter 7

Professor at the University of Manchester

In March 1929, to Hartree's great surprise, the University of Manchester offered him the Professorship that Edward A Milne left vacant when he went to Oxford. This was the Beyer Professorship, Chair of Applied Mathematics. It was a remarkable advancement from Demonstrator to chaired Professor! Many activities – teaching, research supervision, professional service – would have priority over his own research, not to mention the responsibilities of a growing family.

In the early 1930's there were just two professors at the University of Manchester responsible for the Honors program in Mathematics – Professor Mordell who held the Fielden Chair of Pure Mathematics and was the Head of the Department, and Hartree in Applied Mathematics. Jack Howlett [1] recalled that in 1932, about 25 students (approximately equally divided between men and women) started the session, though the number was reduced to a dozen after the outcome of the Part I examination at the end of the year. Those not allowed into the second year were moved to the Pass Degree program. Both Hartree and Mordell took their Honors students very seriously indeed and made a point of lecturing to them in their first year, and again in their third year. Both were first class lecturers.

Howlett got to know Hartree quite well in the student-Professor Relation. He was struck by Hartree's approachability and by his evident concern for his students; he was always ready to go to any amount of trouble to make sure that everyone really understood the idea being put across. When Howlett got his degree (a First), Hartree provided invaluable help in getting a job. This was in Lancashire in 1935, and times were hard. He had read mathematics because he liked the subject but now he was wondering how he should earn a living. He talked things over with Hartree who said he "knew someone in the London, Midland and Scottish Railway (LMS)" who might be interested in a young mathematics graduate. The "someone" turned out to be one of the Vice-Presidents, no less. The end result was that Howlett was offered a job in the Research Department that LMS was about to set up. Howlett claimed this position was very good training since the engineers, for whom he was expected to perform calculations, actually wanted numbers and could not be put off with things like existence theorems.

Mrs Hartree too acknowledged that her husband was a devoted teacher. When Hartree went to Manchester he had done little lecturing, and found it difficult to get much research done in the first three years. He put a great deal of effort into his teaching, never lecturing from old notes, preparing them over and over again each year,

Hartree supervised a number of MSc theses, both in mathematics and physics. Murial Black worked with Hartree on atomic structure calculations (see Chapter 5). Several others were in conjunction with the differential analyzer which will be described in the next Chapter. But he also had two PhD students both of whom went on to have distinguished careers. The first was Arthur Porter who obtained his BSc with first class honors in physics in 1933. He remained for an MSc in 1934, PhD in 1936, and continued as assistant lecturer for a year still collaborating with Hartree. The other was a mathematics student, Oscar Buneman. Buneman [2] had been born in Milan on September 28, 1913 of German parents and began his education in Germany. In 1934 he was imprisoned by the Nazis for political resistance. Upon his release a year later, he came to Manchester University to work with Hartree on the study of nonlinear differential equations.

Manchester kept Hartree rather busy. Upon his arrival, he was invited to give talks to lay audiences, speaking as a physicist. In a letter to Appleton (October 12, 1930) he wrote, "I am not used to such – yet – and in preparing them find I have to take a good bit of trouble ..." Clearly he took these as seriously as his lectures to students. He went on to say he admired Appleton's *Wireless Echoes* lecture given in Bristol and wondered whether it was as easy for Appleton to prepare as it sounded.

In another letter to Appleton (April 4, 1933), he expressed relief at NOT getting a paper to referee. At the same time, it was clear he took such a task extremely seriously. He mentioned having spent half of Christmas vacation refereeing two papers sent to him by the Royal Society – both wrong but it took a long time to find out where they were wrong. After mentioning this, he then went on to suggest a possible referee for the paper Appleton was trying to get reviewed.

He also referred to being asked to serve on a committee about propagation of radio waves. He said that "if it means many visits to London in term time, I definitely cannot manage it. As far as I can see at present, I'll have more lectures and classwork next year than I have had this year, so I will be more tied, and this term it has been difficult to get off any time." Ultimately he talked about some research ideas he would like to pursue.

Mrs Hartree said her husband was not at all fond of Committee work and other administrative functions. At the Annual Meeting of the Mathematical Association, held 8 January 1935, Hartree gave a brief report *The bearing of statistical and quantum mechanics on school work*. Apparently, it was a topic assigned to him. He stated "When first asked to speak on it, my impression was that, as the phrase goes, the answer was a lemon." He continued by saying he thought statistical mechanics had no bearing on school work in mathematics. He thought quantum mechanics was not practical because it would be impossible to go very far, but went on to describe how, a discussion of the physics of "very small" could be interesting. He concluded that "in pointing out the underlying assumptions (of quantum mechanics) that the bearing of quantum mechanics on school work, if any, depends."

Research

As Professor, Hartree was also responsible for promoting research. When he heard that Bohr intended to visit Cambridge, he immediately wrote (April 2, 1930) to invite him to Manchester. Bragg and others would be very pleased to see him. Furthermore, the University would be celebrating the Jubilee of its foundation on May 23-24. Also, Nevill Mott was there as a lecturer in Theoretical Physics, in a position corresponding to the one Bohr held when he was in Manchester. Hartree informed Bohr:

> I remembered discussing him and his work with you shortly before we left Copenhagen, and ... we both felt rather doubtful about his attitude to his work and about whether he would do well in the future as a theoretical physicist. I think he has improved a great deal since then – largely as a result of his visit to Copenhagen and I feel no such doubts now.

Mott had been a great success and both Bragg and Hartree were sorry that he was returning to Cambridge.

Bohr came to Manchester towards the end of May and met many old friends. He told his listeners of a Danish paper in which he had discussed biological and psychological problems from the point of view of what could and could not be observed and described. Hartree requested a spare copy saying his Danish was good enough to follow what had been written.

It was Bohr's custom to organize a conference at his Institute once or twice a year. In 1933 Hartree got invited with an offer of being put up as a guest of the Institute, but could not attend. He was pleased, however, to inform Bohr that the Theoretical Physics at Manchester would be stronger the next year (September 10, 1933).

> We will be enough for a Theoretical Physics Colloquium, which I have wanted to start before, but for which there have not really been enough people who are not rather easily frightened by the mathematics!

Both Hans Bethe and Rudolf Peierls were coming to work with Bragg and Bertha Swirles was returning to the Mathematics staff:

In 1933 Hitler had been appointed chancellor of Germany and by April 7, 1933 prohibited those of "non-Aryan descent" from working at state-run Universities. The speed of events took many by surprise and most Jewish academics fled the country. American immigration policy was cautious and, as a result, many, such as Hans Bethe, migrated first briefly to Britain and then to the USA. In 1932 Peierls had a Rockefeller Fellowship and elected to spend the latter part at Cambridge, which is where he was when turmoil in Germany started.

Another such Jewish refugee was Fritz London who obtained an Imperial Chemical Industries (ICI) fellowship for work at Oxford on superconductivity. Learning of his presence at Oxford, Hartree (June 1, 1934) sent an invitation to him to stay with them for a few days. He mentioned their small physics colloquium, that Peierls would be giving a talk on his work on the positron, and that a second meeting would be nice if London could tell them something about what he has been working on. London was delighted to hear from a "friend," claimed he had not done much but would be willing to participate in scientific life in Manchester for a few days. The efficiency of the Royal mail system is astounding – the following day Hartree responded with instructions on how to meet at the train station.

The visit was more than London had bargained for. During discussions with the experts, Peierls and Bethe had challenged the way in which he determined the number of superconducting electrons, preventing him from giving his planned talk. Upon his return to Oxford, London wrote Hartree to say how well he felt with the Hartree family and thanked them for their kindness (June 8, 1934):

> I have learnt very much in those days of rather continuous discussions; though the criticism of my German colleagues was sometimes pretty hard, it was at least instructive. Of course, scarcely was I in the train when I meditated on the objection by which I was "murdered" by Bethe and Peierls, and, near Barrington, "rising from the dead" saw I was quite right.

Bethe and Peierls wrote a letter of apology, signing it as "pseudo murderers" [3] to which London replied "Murderers, I have not died at all!" Hartree too wrote reiterating that Wigner also thought London's idea was correct, and expressing regret that he had not been able to tell them more about his work.

London's ICI Fellowship was terminated after three years, sooner than expected, and London was again searching for a position. He was hoping for something permanent. Hartree did his best, but all he could do was mention an opening Bragg had and that a fellowship was vacant. By that time London had an offer for an "almost" permanent position at the Institut Henri Poincaré in Paris which he felt compelled to accept.

Life in Manchester

The period in Manchester was a creative period in the lives of Douglas and Elaine Hartree – making friends, growing family, expanding research ideas. They bought a large, Victorian semi-detached, three-story house at 1 Didsbury Park, about five miles from the city center and nearly in the country, yet still on a direct bus route to the University. It had two cellars, one of which was converted into an air-raid shelter during World War II. On the ground floor was a drawing room, extended for two pianos, a dining room, pantry, kitchen, and scullery (a room next to the kitchen for washing and storing dishes, washing vegetables, and similar tasks). The next floor had four bedrooms, a bathroom, and nursery. On the top floor were a study, guest bedroom, and two maid's rooms. In addition there was a separate garage for a car and a good sized garden. They had two live-in domestics, and a nurse/nanny for young children. Their financial circumstances were quite different from that of William Hartree who had been able to retire at the age of 43. In order to pay for family vacations, Douglas Hartree would take on the task of grading examinations during the summer for extra income.

Hartree was an avid photographer in both black and white and color. The top floor of the house had a dark room where he developed his own film and printed enlargements, often making his own "lantern slides." His daughter, Margaret, remembers many times going to Christmas "lectures" which her father and others gave to children in Manchester. They were lavishly illustrated with "magic lantern" shows that he had prepared. He always stayed to answer all questions the children had, showing endless patience and never "talking down" to them [4].

At Manchester the family had a cat that Hartree was fond of. He often worked at home sitting in a big armchair with the cat asleep on his lap. A writing board and papers were balanced on the chair, over the cat, with a Brunsviga on the right arm, and a calculation progressing. The sound of calculators grinding away noisily was common at both Didsbury Park and Bentley Road where his father now lived. The family referred to them as "crashers." Computing was accepted as part of family life. A family joke was "Daddy has lost a factor of 2." When this happened, he would go off on Sundays or holidays and climb a hill to be by himself. Upon his return, the children would ask "Did you find it?" There were universal groans when he said no [5]. Figure 7.1 shows Hartree with his Brunsviga. It is the picture used by the Royal Society for his *Biographical Memoirs*.

According to Rudolf Ernst Peierls in his book *Bird of Passage* [6]:

> In 1933 Manchester was hardly an attractive city. The buildings had been erected mostly during the Victorian period and were in poor taste, and there were many slum areas. The new part where we lived, consisted of cheap houses put up by speculative builders. Most of the older houses were black with soot, so even the few attractive buildings in the centre were not easily distinguished. ...
>
> All this was aggravated by fogs, of which we got a good supply that winter ... But against all these drawbacks the warm and friendly nature of the local population made the city a pleasant place in which to live.

One of the people whose company Peierls enjoyed was that of Douglas Hartree. He said:

> Hartree was a chubby man with the face of an overgrown schoolboy. He was rather shy, and in mixed company his red face would acquire a darker hue before he managed an utterance.

On one occasion, during a discussion on education, Mrs Hartree praised coeducational institutions and Bedales, in particular, for increasing her husband's confidence in social intercourse. Peierls found it difficult to imagine

Fig. 7.1 Douglas Hartree with a mechanical Brunsviga calculator, about 1935.

what Hartree would have been like otherwise. He said the Hartrees were kind and helped with advice and information where they could. The conclusion arrived at from the remark concerning Hartree in mixed company is unexpected. Hartree was not one to participate in "small talk", more common in social settings. Certainly, in scientific work, he related well with women.

Manchester was a railway city. This seems to have provided entertainment for the family as Hartree wrote in a letter (September 29, 1930) to Appleton:

We had a brother and cousin of my wife's staying with us during the week of the Liverpool and Manchester Railway Centenary celebrations at Liverpool, and took them over; also our daughter of 5 1/2 years – good show, much appreciated by all. There were about 15 locos and 20 items of rolling stock in the open in a spacious field, and a train (loco largely original) and railwaymen in the style of the 1830's on which one could ride. The LNER high-pressure compound was there among other things, and I was interested to see it. Some of the locos are here in Manchester this week, and we took the young son of 3 years to see them – great thrill, both to see them and to get square with his sister!

In February, 1931 another son was born to them – they named him John Richard Hartree. In a letter to Waller, Hartree writes that he "arrived about a fortnight before he was expected, and nothing was quite ready, so he has given me rather a lot extra to do and think about ...."

Both of the Hartrees were fond of Manchester. They particularly valued the fact that they had broad contacts and not only with academics as in Cambridge. Mrs Hartree in later years mentioned that they had been glad to raise their children in a community like Manchester rather than in the strictly academic environment of Cambridge. Hartree himself had not found quite as wide a range of scientific life there, but frequent trips to Cambridge, meetings of the Royal Society, and so on, had kept him in touch. Actually, as we will see in the next chapter, Manchester did provide him with some opportunities he might not have had at Cambridge and when the time came to leave, he knew he would miss them.

Musical and social life

During their stay at Manchester, both Hartrees continued with their musical life. According to their son, Richard, Hartree had no great facility for playing either the piano or the clarinet but music had a spiritual, almost religious, aspect for him, bringing him great joy and deep satisfaction. He served on the Council of the Royal Manchester School of Music and as Vice-President of the Manchester University Music Club.

From 1936-46, Hartree served as Dean of the Faculty of Music. In his obituaries it was sometimes stated that Hartree founded the Faculty of Music, in others that he founded the Department of Music.

In fact, the Faculty of Music was established in 1903 when the Victoria University of Manchester was constituted. The Faculty of Music was unusual in that, for many years, it consisted of only the Department of Music.

Thus there was no real distinction between the "Faculty" or "Department". At the same time, during a period when Departments were run by the Professor, Music had no Professor, so the arrangement was even more unusual and informal. In 1935 a new lecturer, Humphrey Proctor-Gregg was appointed who undertook a major reorganization of the teaching of music at the University. In doing so, Procter-Gregg gave the Faculty/Department of Music a much stronger identity. Hartree undoubtedly was involved in this re-organization and strongly supported Procter-Gregg. Thus it is correct to say that during his period as Dean, the Faculty/Department of Music was revitalized to the point where, in 1954, the first Professor of Music was appointed.

Hartree must have been pleased when, in 1937, his friend, Patrick Blackett, assumed the position of Langworthy Professor of Physics at Manchester University. A few days after Blackett's arrival, Bernard Lovell (A radio astronomer who later became the leader of a team that built the world's largest steerable radio-telescope at Jodrell Bank) received a letter from Blackett, indicating that Lovell might be interested in starting a cosmic-ray experiment. Apparently Hartree had suggest Lovell to Blackett. In his book, *Astonomer by Chance* [7], Lovell writes:

> Douglas Hartree was Professor of Mathematics in the university. He and his wife, Elaine, were prominent members of Manchester society and accomplished musicians. It fact, it was rumoured in the university no one would be appointed to Hartree's staff unless he could perform Bach adequately. He had been involved in my appointment, and I was soon summoned to his house for a musical evening during which I was to play a Bach prelude and fugue on their Steinway concert grand. In this way I became involved with the Hartrees, who were immensely kind when I arrived a year later with my young bride, introducing her to outstandingly friendly groups in the university and revealing to us both that the city had attributes more valuable than the superficial rain and fogs that confronted us.

According to Lovell, Blackett organized a conference at Manchester to discuss the existence of the meson. Heisenberg was one of the attendees and stayed with the Hartrees. When he left, he gave Elaine Hartree a rolled package which she, for a moment, thought might be Nazi propaganda, but it turned out to be a fine copy of an autographed score of one of Beethoven's last sonatas that he had performed excellently during his stay.

The Hartree family Visitors' book contains the names of close friends – Patrick and Constanza Blackett, Alex and Alison Todd, Charles and

Katherine Darwin – and others, notable in Hartree's professional world, namely Nevill and Ruth Mott, John and Helen Slater, Niels and Margethe Bohr, Edward A Milne, Walter Heitler, Vannevar and Phoebe Bush, John E Lennard-Jones, Harold and Katherine Hazen, Subrahmanyan Chandrasekar, John D Cockcroft, Maurice V Wilkes, Ivar and Irene Waller, Werner Heisenberg, Bernard and Joyce Lovell, and Egil A Hylleraas. Elaine Hartree played an important role in establishing and maintaining contacts, complementing Hartree effectively.

Robert Bruce Lindsay, a visitor from America

In 1930, Lindsay moved to his Alma Mater, Brown University, where he remained for the rest of his professional life. His research was moving in the direction of acoustics, but he was not yet confident of directing theses in the area. With Hartree's advice, he was comfortable directing students in quantum mechanics. In his *Intellectual autobiography* [8] he said:

> I had corresponded with Hartree regularly, who very generously gave me complete directions about carrying out the numerical integration of Schrödinger's equation for the polyelectronic atoms, and suggested that I might guide some graduate students through such calculations. He wrote very voluminously about his own calculations and gave many useful hints.

For his first sabbatical year, Lindsay applied for a Guggenheim Fellowship to spend 1936-37 working with Hartree at Manchester. Hartree was delighted and responded with a list of possible topics they could work on: extensions of SCF to include exchange and relativistic effects, configuration interaction, or anything Lindsay was interested in. But the Fellowship was denied. Lindsay and his wife, decided instead to spend January - August 1937 abroad out of personal funds. Arrangements were made to work at Imperial College of Science and Technology of the University of London. Douglas Hartree arranged for an introduction to an acoustical consultant for Lindsay.

The Hartrees had invited them to Manchester and they spent a week there in May 1937. Mrs Hartree took them on trips to the surrounding country. Douglas became aware of their son Bob's interest in railroads and arranged a special excursion for Bob, Lindsay, Hartree's son Oliver, and Bragg's two sons to the locomotive works of the LNER at Gorton. This was a fascinating experience, especially for son Bob.

Lindsay said that the visit to Manchester was made particularly pleasant by the chance to meet so many of the staff who later became quite distinguished. In addition to Bragg, there was Polyani the well known Physical Chemist, and Tolansky whose work on optical interference made him famous. He met Bertha Swirles, then lecturer in Mathematics, "who made one of the earliest (probably the earliest) relativistic self-consistent field calculations under the direction of Professor Hartree."

With Hartree's advice, a number of PhD theses followed this visit: Robert Lee Mooney (1938) *A self-consistent field for doubly ionized chromium*; Clayton Roy Lewis (1939) *Collision cross-sections of some singly-excited states of helium using Hartree wave functions*; William Jacque Yost (1940) *Self-consistent fields for doubly ionized magnesium*, and Arthur O. Williams (1940) *A relativistic self-consistent field for Cu^+*. By 1940, Lindsay was well established in acoustical research and directed no further theses in atomic structure work though Hartree and Lindsay remained friends.

Notes and References

[1] Letter from Jack Howlett in May, 1999
[2] A tribute to his memory was published by Buneman R, Barker R J, Peratt A L, Brect S H, Langdon A B, and Lewis H R 1994 *IEEE Transactions on Plasma Science* **22** 22
[3] Gavroglu K 1995 *Fritz London: a scientific biography* (Cambridge University Press)
[4] Booth, Margaret Hartree 1986 *Douglas Rayner Hartree: a personal memoir* Christ's College, Cambridge
[5] From notes of an Oral Interview with Elaine Hartree, May 12, 1963, American Institute of Physics, Center for History of Physics, Niels Bohr Library, College Park, MD
[6] Peierls R E Sir 1985 *Bird of Passage* (Princeton, NJ: Princeton University Press)
[7] Lovell B 1990 *Astronomer by Chance* (New York: Basic Books, Inc.)
[8] Lindsay R B 1962 *Intellectual Autobiography*, Niels Bohr Library, Center for History of Physics, American Institute of Physics, College Park, MD

Chapter 8

The differential analyzer

In November, 1931 when Slater conveyed to Hartree news about what was going on at MIT he mentioned:

> Then another very useful thing is the differential analyzer which Professor Bush of the Electrical Engineering Department has built. He has been interested in computing for years, and this is his latest, most astonishing device. It is a machine for solving differential or integral equations, and it handles one dimensional wave functions very nicely.

Hartree had already read a brief account of the machine in the latest edition of the Encyclopedia Britannica, and the solution of the radial equation at once occurred to him as an application. What concerned him though was that the model was stated as having an accuracy of only 1-2% which he did not believe would be sufficient for his work. Accuracy was crucial to him. Most of his work was done to 5 significant figures in the radial function. But he was intrigued, and requested a copy of Bush's paper[1] and went on to say (December 21, 1931):

> I am extremely interested to hear of Professor Bush's latest machine, and the possibility of using it on atomic calculations; it alters the technical computing problem considerably, more than I realized .... If you can do in 10 minutes what takes me 2-3 hours, and with adequate accuracy, it seems to bring the solution of Fock equations within the range of possibility.

He talked about his experience with an integraph device during WW I. It was a beautiful instrument, he said, but he returned to numerical methods in the end, primarily for certainty in accuracy.

In the face of much skepticism from the faculty, Vannevar Bush had qualified, after only one year, for a doctorate in electrical engineering in

1916 from both Harvard *and* MIT. He was an exceptional engineer. He returned to MIT in 1919 as Associate Professor of Electrical Power Transmission and in 1923 became Vice-President and Dean of the Engineering School. There he and his graduate students were working on integrators. One of them was Harold Hazen (1901 - 1980) who came up with the idea of "coupling" integrators. He sketched his idea of a mechanical disk-and-wheel integrator and its incorporation into a machine [2]. Bush quickly recognized the significance of the idea and proceeded to build a six integrator machine with Hazen contributing to the design. In his later differential analyzer work, Bush had the invaluable assistance of Samuel H Caldwell. In fact, together they tackled the solution of the Thomas-Fermi equation as their first problem because it brought out certain interesting points in connection with the operation of the machine [3]. Like the central field problem, there were two point boundary conditions, the differential equation was non-linear and involved a singularity at the origin.

Hartree made arrangements to visit Bush at MIT in the summer of 1932 [4]. To become thoroughly familiar with the operation of the machine he made a second visit in 1933 during which he attempted to solve the self-consistent field equations for mercury (Hg). It was here that he first introduced the independent variable $\rho = \log r$ for automatic computations. Several self-consistent field iterations were completed, but the $(4f)^{14}$ group was, as Hartree stated, "inconveniently sensitive" and the task was not completed before his visit was over.

But Hartree was not only interested in the self-consistent field or Fock equations, he was intrigued by all forms of differential equations. In a paper read at the Annual Meeting of the Mathematical Association in January, 1938, Hartree discussed the *The Mechanical Integration of Differential Equations*. Here he made clear that differential equations without formal solution occur in a wide range of applications of mathematics to pure and applied science, particularly in physics and electrical engineering. A mechanical means of solving differential equations that was rapid and accurate enough for practical purposes, easily applied to a wide range of problems, was an important technical advance. The use of mechanical integrators had been suggested by Kelvin already in 1876 and various special-purpose devices had been constructed at various times. But the first device to meet some genuine needs, in Hartree's opinion, was the one designed and built by Dr Vannevar Bush that Bush called a *Differential Analyzer* [5]. In Hartree's talk at the Mathematical Association meeting, he remarked:

The name of this machine, by the way, seems to be scarcely appropriate, for the machine neither differentiates nor analyses, but, much more nearly, carries out the inverse of each of these operations: however it is Dr. Bush's child, and I think he has the right to christen it.

In order to understand Hartree's many contributions to the design and application of differential analyzers, it is necessary to have some basic understanding of underlying concepts.

A distinguishing component of the differential analyzer was the integrator. Consider a rotating, horizontal disk on a vertical *driving shaft*, supported on bearings to a movable carriage. On the disk is an *integrator wheel* at a fixed plane in space with a horizontal output shaft. Thus the position of the wheel from the center of the disk may vary. As a motor rotates the driving shaft an amount dx, friction will rotate the output shaft an amount proportional to ydx as depicted in Figure 8.1 and thus, over time, the cumulative rotations of the output shaft is $\int y(x)dx$. All this is achieved with a motor, shafts, and variable gears. However, it is important that slippage of the integrating wheel be minimized and for this purpose Bush and his group designed a torque amplifier which permitted the shafts and gears to respond to the smallest of forces.

Other components were input tables. To solve the differential equation

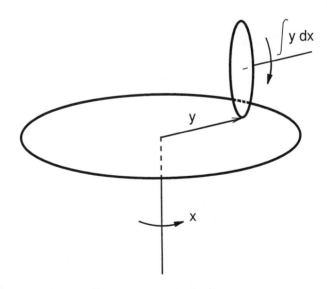

Fig. 8.1 Diagram depicting an integrator

$dy/dx = f(x)$, the function $f(x)$ was represented on a graph placed on the input table. As the motor controlled the rotations representing the variable x, a bridge over the table moved across the table. The bridge supported a carriage with a pointer that could be displaced manually along the bridge by a handle, so that the pointer remained on the curve. A drive from the handle then controlled the displacement of the horizontal disk.

Also important were output tables to display the results. These were similar to input tables except that a pen or pencil replaced the pointer and was driven in both x- and y-directions by the machine. An alternative method could be the recording of revolution counters using a specially designed camera. Adding units were mechanisms connecting three shafts so that the rotation of one was the sum of the rotations of the other two. Multiplication by a constant was effected by connecting the appropriate shaft to a second shaft through gearing, provided that the constant could be expressed to adequate accuracy as a ratio of convenient integers. Thus a differential analyzer consisted of a collection of integrators, adders, multipliers, input tables and output tables. These components could be connected in various ways through shafts and a variety of gears, so as to represent the solution of a problem.

To use the analyzer, it was necessary to first express the solution of the problem in terms of a schematic, a notation which had been proposed by Bush. An example that Hartree himself used is given in Figure 8.2. The symbols on top represent integrators with their inputs and output and others show summation. This example shows that solutions could be obtained in ways not considered by other methods and that the special features of the differential analyzer could be used to advantage.

The equation he considered was

$$\frac{d^2y}{dx^2} - y\cos(x+y) + \left(\frac{dy}{dx}\right)^2 = 0.$$

The function $\cos(x+y)$ could be obtained from the solution of a second order differential equation, requiring two integrators, one for each order of differentiation. But first the equation for $y(x)$ needed to be rewritten as a first-order equation, namely

$$\frac{dy}{dx} = \int y\cos(x+y)dx - \int \frac{dy}{dx}dy.$$

Note that the second integral on the right hand side is now an integral with respect to y. It was a special feature of a differential analyzer that

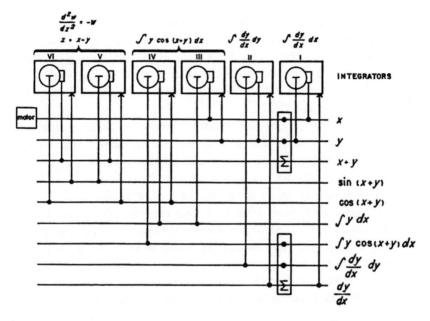

Fig. 8.2 Schematic for the solution of the differential equation $\frac{d^2y}{dx^2} - y\cos(x+y) + \left(\frac{dy}{dx}\right)^2 = 0$. (From Hartree D R 1949 *Calculating Instruments and Machines* (Urbana: University of Illinois Press)).

the integration variable could so readily be changed. The solution of the problem first required arranging the connections between shafts so that the rotations of the shaft represented the equation being solved, as indicated by the schematic in Figure 8.2. The initial conditions of the problem were usually inserted in the form of initial displacements of the integrators. In a typical problem it took half a day to set up the machine and one solution could be obtained in perhaps 15 minutes of machine time. The use of the machine was not particularly fast if only one solution of a simple equation was required, but it was efficient if a long series of solutions with different values of various parameters were required. It is also interesting to note that the motor driving the independent variable did not have to rotate at constant speed. In fact, it was useful to slow down the motion in order to carefully follow an input curve, a task that needed to be done manually. For the evaluation of the final solution, a variety of scaling factors also needed to be considered.

On seeing a photograph of Bush's differential analyzer, Hartree's first impression had been that "someone had been enjoying himself with a large Meccano set." Meccano was a children's construction toy widely available in England and purchased by many parents before World War II, though it was not considered a cheap toy. A kit included wheels, tires, gear wheels, shafts, and other mechanical items as well as girders, struts, brackets and nuts and bolts to make structures. Meccano had provided Hartree with many happy hours as a child.

Upon his return to England, Hartree decided to build a differential analyzer of his own, first building a demonstration model using Meccano before seeking funding for a full-size model. In this work, he was joined by his student, Arthur Porter.

The model differential analyzer was built very largely from standard Meccano parts, with some additional long screws and a special motor made for this purpose by Meccano. Some additional helical gears, sprocket wheels, and chains were purchased from Bonds in London. Initially it had three integrators but a fourth was added soon afterwards. Single torque amplifiers were used initially with an amplification factor of about 80. Later a two-stage torque amplifier was introduced with an amplification factor of roughly 2,000 that improved the accuracy of operation. The total cost of the model machine parts was estimated to be only £20, which was a remarkably small sum, even in the economic depression of the time. A paper describing the model differential analyzer was published by D R Hartree and A Porter in *Memoirs and Proceedings of the Manchester Literary and Philosophical Society* in 1935.

News of this unusual machine first appeared in the local *Manchester Guardian* on January 24, 1934. The article *A diligent machine* made the point that, the "adroit and accomplished" machine doing the "donkey's work", might lead to better scientists and more of them. A second article, in the same issue, written by the Scientific Correspondent, included a description of the machine and reflected Hartree's views. It was estimated that the machine could save 60-90 % of the calculator's time, that:

> Original numerical calculations must always be done, however laborious, by their discoverers. Hence investigators of the first class often either shun such calculations to the retardation of their subject, or they devote too much of their energy to them.

News was also soon reported to Meccano enthusiasts. The June, 1934 issue of the *Meccano Magazine*, page 441 featured an article by the Editor, entitled *Meccano Aids Scientific Research* in which it was stated:

The Differential Analyzer constructed by Professor Hartree illustrates one of the most valuable features of the Meccano system, namely, its infinite elasticity. The original American machine is built up of a number of units, each playing its own part in the intricate process of solving the difficult mathematical problems with which it is designed to cope; and its range can be extended almost indefinitely by the addition of further units. The elasticity of the Meccano has made it possible for the model to be built up and extended step by step to adapt it to new problems. The results that Professor Hartree has already achieved with the model in his research work are sufficient in themselves to justify the claim that it represents the most remarkable scientific application of Meccano parts that has yet been made.

An article, *Machine Solves Mathematical Problems*, followed showing the picture of the differential analyzer, with D R Hartree and A Porter (also shown here in Figure 8.3). It included some very clear pictures of the input and output tables, the model in its earlier form, and one of the integrators of the model.

The above publication provided only meager details on how the machine was constructed and how it was to be used. In the *GMM Series of Modern Supermodels*, published by the Meccanoman's Club, Henley-on-Thames, England, December, 1967, detailed instructions were provided, even though the differential analyzer had long been superseded by electronic devices.

It is not surprising, that with this response, a number of other machines were built, either of standard parts or with some parts built in a workshop. These included a machine built at the University of Cambridge by Lennard-Jones and at Belfast by Massey [6]. There were other smaller versions. Hartree mentioned one model built at Macclesfield Grammar School by a teacher, R Stone, and some of his VI'th Form boys. R E Beard [7] built a small machine of six integrating units and four input tables for actuarial work.

The first application of the Meccano model machine was by Porter [8; 9]. He used it to obtain preliminary self-consistent field (SCF) wave functions for chromium in the $3d^4$ and $3d^6$ configurations. A feature of the radial equation is that the solutions may oscillate fairly rapidly near the nucleus and then decay exponentially. In the numerical integration of the SCF equations, it was common to start the integrations with a small step and periodically double the step-size as the radial function began to vary less rapidly. In using the differential analyzer, it also was desirable to keep solutions from oscillating too rapidly in order to maintain accuracy. An idea, suggested by Hartree in his work on mercury (Hg) at MIT, was to change the independent variable in a fashion which simulated an increasing step size. With $\rho = \log r$, we have $dr = r d\rho$. Consequently as ρ changes uniformly, the values of r change slowly near the origin but they change rapidly at large r. In order to keep the form of the differential equation that of a second-order differential equation without a first derivative, it was also necessary to change the dependent variable to $Pr^{-1/2}$ so that the final radial equation to be solved was

$$\frac{d^2(Pr^{-1/2})}{d\rho^2} + \left[2Z_p r - \varepsilon r^2 - (l+\frac{1}{2})^2\right] Pr^{-1/2} = 0, \qquad (8.1)$$

where Z_p/r was the potential at radius r. Porter estimated that the integrations from the differential analyzer might be in error by 5 per cent. It should be mentioned that this equation is an eigenvalue problem in that ε needs to be determined in such a way that the solution is not only zero at the origin, but decays exponentially at large r. This would require several runs itself before the boundary conditions were satisfied. There also were 7 equations to be solved iteratively using the self-consistent field method. Not exactly a trivial task at the time.

The Manchester differential analyzer

With the success of his model analyzer, funding was obtained and a full size machine was constructed at the University of Manchester to be housed in the basement of the Physics building. Svein Rosseland, who had plans to build a differential analyzer at the University of Oslo, Norway, had visited Bush in the early Spring of 1933. During this visit he produced complete and thorough documentation of the machine [10]. Bush freely gave these drawings to Hartree along with suggestions for improvements and a great deal of advice based on his own experience. The machine was built in two sections, the first including four integrators, two input tables, an output table, and a special recording camera. Because of Hartree's particular interest in the study of time-lag in control systems (see Chapter 9), a special input table was constructed, essentially a combination of an output and an input table. It allowed the result usually sent to an output table to be redirected, with a time delay, as an input variable. The second section also had four integrators and additional input tables. The machine occupied the whole floor space of a room of about 25 ft by 20 ft.

The construction of the machine was entrusted to the Metropolitan-

Vickers Electrical Company Ltd, and was built under the supervision of A P M Fleming (1881-1960), then Manager of Research, with the assistance of many other members of the firm. Later he was responsible for directing various problems to Hartree for him to solve on the analyzer. The unique high-speed camera for recording numerical results was designed and constructed by Messrs Newman & Guardia [14].

Hartree's funding came from a private source. Mr Robert McDougall (1871 - 1938) had been brought up in the family flour mills business, and later became a director of companies in the flour milling industry. He made many gifts to charity. He was the unpaid deputy treasurer of the University of Manchester. He generously donated funds for the construction of the machine, amounting to £6,000 for the first section and a similar amount for the second section. The machine was dedicated on March 27, 1935 at a distinguished gathering presided over by the Lord Crawford, Chancellor of the University. Hartree gave a speech mentioning the wide range of problems for which it could be used, paying tribute to the donor for his support of this powerful research tool, thanking Bush for his friendly cooperation in its design, and expressing his appreciation to various members of the firm who had been involved in its design and construction. After speeches from various dignitaries, the gathering descended to the basement of the physics building to see a demonstration of the machine in operation. According to Peierls, the day before the official opening ceremony, the technicians gave the machine a fresh coat of paint. Unfortunately, they also painted the moving parts, so the machine stopped functioning. Hartree and his students spent the night removing the paint from all parts where it was not wanted [12].

Again, the *Manchester Guardian*, on March 27, 1935, was the first to announce this new, marvellous machine. Already job security was an issue, a comparison being made with calculating machines that were revolutionizing accounting in banking and business, but it was argued that there was a difference:

> The scientist who finds a calculation too laborious can neglect it and turn to one that is more promising whereas accounting has to be done by someone, ...

Hartree also demonstrated the machine to journalists. A picture of the event appeared in the *Manchester Guardian* on March 28, 1935 and is shown in Figure 8.4. But this time, the news also went further afield. The *Times* (London) featured an article about the "ingenuous machine for

Fig. 8.3   Douglas Hartree and Arthur Porter viewing the Meccano differential analyzer, 1935.

Manchester" (March 28, 1935). The *Observer* (March 31, 1935), in its section *At Random*, referred to the machine as getting dangerously close to a robot and wondered what chance an ordinary man with a multiplication table had in competition with the machine. Articles also appeared in *World Power* [13], *The Engineer* [14], *Engineering* [15], and in *Nature* [16].

In addition, Hartree sent short letters to *Nature* and *The Mathematical Gazette*. A few years later, Hartree and Nuttall, a member of the Research Department of Metropolitan Vickers Electrical Co, Ltd and an Associate Member of the Institute of Electrical Engineers, sent a report to the *Journal of the Institute of Electrical Engineers*. Thus the scientific, mathematical, and electrical engineering community was alerted to his achievement.

Bush and Hartree remained in close contact over the next 15 or so years. Both played important roles in World War II and, when Hartree was back at Cambridge, he arranged for Bush to receive an Honorary degree from Cambridge University in 1950. Figure 8.5 shows the two men in their academic gowns just prior to the ceremony. At that time, Bush was President of the Carnegie Corporation, a position he held from 1939-1955. During World War II, he had been Director of the Office of Scientific Research and Development having been appointed by President Franklin D Roosevelt. His report to the President, entitled *Science, the Endless Frontier*, secured the establishment of the National Science Foundation.

Early Applications

Among the early papers to appear was a study of the effect of space-charge on the secondary current in a triode. The theory was developed by Myers, Hartree, and Porter. This work prepared Hartree for later magnetron research during the war.

When electrons strike a surface with sufficient velocity, secondary electrons are emitted by it. In this work, a triode is considered, in which the anode receives electrons from the cathode and emits secondary electrons. These are collected by a grid between the cathode and the anode, maintained at a higher potential than the anode, resulting in a cloud of electrons between the cathode and the anode. An axially symmetric configuration was used. Of interest was the "space-charge limitation" of secondary current. In the limiting case of no space charge, all secondary electrons emitted by the anode would reach the grid. Any appreciable reduction must then be due to space-charge limitation.

Fig. 8.4 Picture of Professor Douglas Hartree demonstrating the Manchester differential analyzer to journalists appearing in the *Manchester Guardian*, Tuesday, March 28, 1935. (Courtesy of Dr Jon Agar and Simone Turchetti, Centre for History of Science, Technology and Medicine of the University of Manchester. Reprinted by permission of the *Manchester Guardian*.)

In order to use the differential analyzer for this problem, it was first necessary to formulate the problem mathematically, in terms of differential equations. These were then reduced to non-dimensional quantities and the equations transformed to their simplest possible form. A schematic was prepared which described the necessary set-up of the differential analyzer. The calculations relied on some initial numerical integrations that were carried out by the Metropolitan-Vickers Electrical Company. A number of cases were considered and for each case, families of solutions were obtained for different parameters. These provided insight into the nature of the solutions of the equations. At that point, comparisons were made with experiment performed by Myers at the Engineering Laboratory at Oxford University. It was concluded that theory agreed with experiment provided the initial energy of the secondary electrons was given a value of about 4 volts. This was taken as an approximate indication of the average initial energy of secondary electrons emitted by a nickel electrode.

A novel application was to the calculation of train running times. This was a practical problem engaging the active interest of organizations concerned with railway transport.

Hartree was known for his interest in trains, both real and model. In 1926, during the General Strike, Hartree helped man the signal box at a level crossing [17]. He enjoyed this immensely and it spurred his interest in railroad signaling the rest of his life. He took his children on innumerable train spotting trips, often accompanied by some of his students. With infinite patience, he taught them to identify engine classes, wheel arrangements, and destinations. On train journeys, he enlivened the trip by teaching them how to calculate the speed of the train by counting telegraph poles, recognizing rail lengths by the noises of the wheels made going over joints, and so on. During vacations, every opportunity was made to visit narrow gauge railways in the vicinity, talking to engine drivers, and signalmen. Thus it is no surprise that, in another application, the differential analyzer was applied to the calculation of train running times.

The problem was a comparatively simple example of integration of a differential equation. Numerical and graphical methods were being used at that time, both tedious and lengthy. The problem was also interesting to Hartree in that the running of a train is subject to speed restrictions, a situation that had no parallel in previous work that had been done on the differential analyzer. After the completion of the work with the assistance of John Ingham, it was learned that a special integrating machine had been built for this problem, very similar to a portion of the differential analyzer.

In Hartree's usual style, he stated the problem as a differential equation for the rate of change of velocity (acceleration) in terms of the tractive force of the engine, T, the masses of the engine and the train (M and m), the resistance of the train and engine (R and r, both a function of the velocity v) and the gradient $1/n$, namely

$$\frac{dv}{dt} = \frac{T - R - r}{M + m} - \frac{g}{n}, \tag{8.2}$$

where g was the gravitational constant. Then he expressed the equation in a more convenient form for the differential analyzer. The solution required a considerable amount of manual intervention. The function $(T-R-r)/(M+m)$ was provided on the input table as a function of v. An assumption was made that the gradient was constant over long stretches so the machine was stopped and the quantity g/n entered manually whenever the gradient changed. It was also assumed that the speed restriction was a function of the distance. This required an analysis of the different cases: one where the masses of the engine and train slowed the acceleration and one where braking was needed. As his example, he chose the run from Rugby to London (Euston).

The paper describing this application, exhibits a detailed knowledge of the running of trains.

Differential analyzers in computing centers

Hartree generously made the analyzer, housed in the basement of the Physics building, available to those who wanted to use it, often providing help, but it was not a general user facility. Yet the differential analyzer was the first tool best utilized as part of a "Computing Center." Hartree's differential analyzer was to have a significant role in the development of the Mathematical Laboratory for research and teaching at Cambridge University [18].

The work of the Theoretical Chemistry Group, led by Lennard-Jones, required the computation of wave functions. By 1935, the complexity of the calculations were limited by the desk calculators being used. When Lennard-Jones became aware of the model differential analyzer, he contacted Hartree, who was happy to give advice and assistance for the construction of a similar machine at Cambridge. The machine was built in 1935 by J B Bratt under the direction of Lennard-Jones. It was a useful computing tool. The day-to-day running of the machine was delegated to Bratt. In March 1936, Maurice V Wilkes, then associated with the Cavendish Laboratory Radio Group, heard of the machine and saw immediately how the device could solve some of the equations that arose in his research. By the end of 1936, Wilkes replaced Bratt in the job of maintaining the analyzer.

The model differential analyzer demonstrated the potential value of a larger, more accurate machine. Lennard-Jones began to develop support for such a machine, not for the Theoretical Chemistry Group but for the benefit of all scientists in Cambridge. He was interested in establishing a computing laboratory, the principal machine of which would be the differential analyzer. Cambridge University gave formal approval on February 20, 1937 for the establishment of a Mathematical Laboratory to be equipped with a new 8-integrator differential analyzer, the model differential analyzer, and other machines including desk calculators. Lennard-Jones was appointed part-time director whereas Wilkes was appointed as University demonstrator, with responsibility for the supervision of the construction of the new differential analyzer. A trip to Manchester was in order. According to Wilkes [19]:

> This trip was the occasion of my first meeting with Professor Hartree who in later years became my close counsellor and friend. He would go to any amount of trouble to help people. On this occasion, not only did he invite me to stay at his home, but he met me at the station and drove me to the University. He introduced me to Arthur Porter, who lent me a Laboratory coat, gave me an Allen key and let me try my hand at making some adjustments to the set-up of the differential analyzer. Setting up a mechanical differential analyzer was not a job for anyone who liked to keep his hands clean; it consisted of arranging in the desired configuration a lot of oily shafts and gears. During my visit, Hartree talked to me about some of the computing that he was doing with a desk machine. In spite of his modest manner, I sensed that here was a level of professionalism in computing far beyond anything I had encountered up to that time.

The Cambridge machine was on order by 1938 but by the time it was delivered in October 1939, World War II had broken out at which time the Laboratory premises and its equipment were leased to the Ministry of Supply. After the war, Wilkes was appointed first Acting Director in 1945, and a year later as Director.

Thus it would seem that at Cambridge, the differential analyzer was seen as a useful research tool that led to the development of a computing center.

On a more modern note, a 1999 issue of the *New Scientist*, in the section *The Last Word*, posed the question under the heading *Toy Story*:

> While toying with a Meccano model, I wondered if any significant inventions or principles owe their discovery to the use of Meccano, Lego, or similar construction toys.

There were several answers, but one came from Justin Hartree:

> My grandfather Douglas Hartree used Meccano to build several large

calculus machines for numerical analysis in the 1940's and 1950's. There is an exhibit in the permanent collection at the Science Museum, London, which features part of one of the inventions. He wrote several books on the subject, some of which contain photographs of the machines.

Seems the differential analyzer is still fascinating to some.

Notes and References

[1] Bush V 1931 *J. Franklin Inst.* **212** 447
[2] Brown G S 1981 *IEEE Annals of the History of Computing* **3** No 1, 4
[3] Bush V and Caldwell S H 1931 *Phys. Rev.* **38** 1898
[4] Arthur Porter is certain Hartree visited Bush in 1932. In a paper submitted by Hartree on August 19, 1934 to *Physical Review*, he reports on the work he started on calculations for Mercury (Hg) "last summer", which implies a second visit in 1933.
[5] In this book, I have adopted American spelling. On the web, useful links are found with this spelling as well as the "differential analyser" spelling.
[6] Massey H S W, Wylie J, Buckingham R A and Sullivan R 1938 *Proc. Roy. Irish Acad.* **45a** 1
[7] Beard R E 1941 *J. Inst. Actuaries* **71** 193
[8] Porter A 1934-5 *Mem. Proc. Manchester Lit. Phil. Soc.* **79** 75
[9] Porter A 2003 *IEEE Annals of the History of Computing* **25** No 2, 86
[10] Holst, Per A 1996 *IEEE Annals of the History of Computing* **18** No 4, 16
[11] Marsh P 1978 *New Scientist* **80** No 1134, 28
[12] Peierls R 1985 *Bird of Passage* (Princeton, NJ: Princeton University Press) p 105
[13] Anon. May 1935 *World Power* **23** No 137
[14] Anon. July 1935 *The Engineer* **160** No 4149, 56; *ibid* No 4150, 82
[15] Anon. July 1935 *Engineering* **140** No 3628, 88
[16] Anon. 1935 *Nature* **135** 535
[17] Booth, Margaret Hartree 1986 *Douglas Rayner Hartree: a personal memoir* Christ's College, Cambridge
[18] Croarken M 1992 *IEEE Annals of the History of Computing* **14** No 4, 10
[19] Wilkes M V 1985 *Memoirs of a Computer Pioneer* (Cambridge: MIT Press)

Fig. 8.5 Douglas Hartree (left) and Vannevar Bush (right) prior to the Honorary Degree Ceremony, Cambrige, England, 1950.

Chapter 9

Control theory and industrial applications

In a letter to Lindsay (October 11, 1936) Hartree confided:

> ...after all, I am interested in other things besides atomic structure computations and actually, since I have had the help of my father in that work ... have been doing less and less of the actual computing work in that field myself; and the differential analyzer has been bringing me into contact with various branches of pure and applied physics with which I was barely acquainted previously, and I am finding the experience both stimulating and interesting.

One of those branches was control theory.

An outcome of the industrial revolution was the vast expansion of manufacturing plants in which there was a need for the control of variables, such as the temperature of reactants or the rotation speed of a component. Governors were invented, such as Watt's governor for steam engines. The book, *Automatic Control: Classical Linear Theory*, edited by George J Thaler, in the series *Benchmark Papers in Electrical Engineering and Computer Science* [1] outlines the development in the field, starting with a discussion of the capabilities and limitations of governors. This volume presents a selection of research papers that have had considerable influence on the development of control theory and covers the period 1868-1950. Both of Hartree's papers on the subject are included among the selected papers. The first was by Arthur Callender, Douglas Hartree, and Arthur Porter, with Hartree the first author on the second paper. In fact, these papers followed one written by Harold Hazen [2] whose contribution to Bush's differential analyzer was the design and construction of the wheel and disk integrators with their torque amplifiers. Hazen was the first to write a definitive paper on feedback controls, which marked the beginning of an era. In time, he and Hartree became great friends. Like Hartree, he was drawn to teaching and for many years served as Dean of the Graduate School at MIT.

The type of control Hazen was concerned with, such as time-operated traffic signals or a thermostat, was somewhat different from the continuous control Hartree and his collaborators were investigating. The effect of "time-lag" was considered in the papers by Hazen as well as those by Hartree, but the case was made that the control of a quantity subject to uncontrolled disturbances was different from the types of control that Hazen had considered. In any event, the two pieces of work were carried out totally independently.

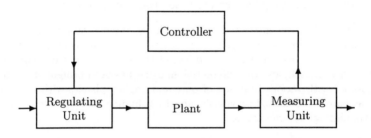

Fig. 9.1 Schematic of a Process Control System

The principal elements in an automatic-process-control system are shown in Figure 9.1. The measuring unit records the value of the quantity to be controlled that depends on plant conditions. The purpose of the controller is to maintain this quantity at a desired value or "control point." The controller compares the measured value with the control point and determines some control action. This action operates the regulating unit to adjust physical quantities that affect the plant operation.

In 1934, Albert Callender, a research chemist who worked in the Research Department of the Alkali Division of Imperial Chemical Industries, approached Hartree with a problem relating to the control of a chemical plant. A long investigation was undertaken. According to Porter [3], Hartree developed the mathematical model, which was then solved in different ways: Porter solved the equations on the differential analyzer and Callender built a model to test the theory. The working model was exhibited by ICI (Alkali) Ltd at the Royal Society Conversazione on May 3, 1935.

The first step was to define the problem in terms of a differential equation. The authors were familiar with the concept of time-lag in servo-mechanisms and this notion was introduced into their analysis. They clarified that a time-lag may be a practical formulation of a process (or an approximation) and not an exact description. For example, an increase in the flow of steam through heating pipes does not immediately increase the temperature of the medium surrounding the pipes because a strict description would need to include the heat transfer problem through the walls of the pipe and through the medium. This might be a slow process that could be modeled instead as time-lag.

They defined $\theta(t)$ to be the deviation of the quantity to be controlled. Then the change in this quantity was expressed as

$$\frac{d\theta(t)}{dt} = D(t) + C(t) - m\theta(t), \tag{9.1}$$

where $D(t)$ was the effect of random (uncontrolled) disturbances, regarded as a given function of t and called the "disturbing function," $C(t)$ was the effect of the operation of control at time t, $-m\theta(t)$ was the inherent effect of the variation of $\theta(t)$ from its zero value. The problem was to study the nature of the control function, $C(t)$, for various types of disturbances.

The assumption was made that the mode of operation was determined by the quantity to be controlled. For a system with a finite time-lag, the function $C(t)$ would depend not on $\theta(t)$ but $\theta(t-T)$, where T is the time-lag. Specified another way, the control at time $t+T$ is determined by $\theta(t)$. From these arguments, they postulated

$$-\dot{C}(t+T) = n_1\theta(t) + n_2\dot{\theta}(t) + n_3\ddot{\theta}(t), \tag{9.2}$$

or, in integrated form,

$$-C(t+T) = n_1 \int_d^t \theta(s)ds + n_2\theta(t) + n_3\dot{\theta}(t). \tag{9.3}$$

In the above, dots refer to differentiation. In later years, the portion of the control law proportional to $\theta(t)$ was called *proportional action*. The part of the control that depended on the values of θ in getting from some point in the past, say $\theta(d)$, to $\theta(t)$ was the integral with respect to time and denoted as the *integral action*. It is the term which takes into account the "history" of the quantity being controlled. The desired control might also depend on the rate of change of θ itself and this was known as the *derivative action*. For a set of parameters, n_1, n_2, and n_3 Equation (9.3) defines a *control law*.

There are two conditions which control must satisfy in order to be practical: it must be stable (without oscillations of increasing amplitude) and it must have quick response. This is determined by the parameters n_1, n_2, and n_3 which determine the amount of proportional, integral, and derivative action. The primary object of their first paper was to make a theoretical study of a class of control systems with time-lag and obtain some guidance as to the optimum values of parameters for the control function. The role of the different parameters was investigated.

Callender, Hartree, and Porter, considered the behavior of the control system for different disturbances and suitable ranges of values of parameters. Heaviside operators were used for an analytic analysis, but also the differential analyzer, for which this problem was fairly well suited. The apparatus could be set up and families of solutions could be obtained by changing the parameter values from one run to the next. In their first paper on *Time-lag in control systems* (1936), they concluded that "a practical method, feasible on the industrial scale, for obtaining automatic control approximately ... is outlined, and two analogous methods are indicated."

In the second paper by Hartree, Porter, Callender and Stevenson (also at the Research Department of ICI), a somewhat more complex control law was investigated. In both papers, the authors are concerned with system performance in parameter space. Their optimization consisted largely in establishing bounds on the parameters. The differential analyzer studies indicated the nature of the response and system sensitivity to parameter variation.

Hartree's formulation of the problem, as given by Equation (1), led naturally to the "derivative" controller. According to Porter [3], this was a new concept in control theory.

In chemical process control, a frequent reference is the paper by J G Zeigler and N B Nichols, *Optimum Settings for Automatic Controllers*, published in 1942 [4]. Both authors were employed at the Taylor Instrument Company in Rochester. The paper was presented at a meeting of The American Society of Mechanical Engineers and was concerned with practical adjustments of automatic controllers. Discussions were carried out without resorting to mathematics, yet they clearly defined and explained the three types of control: proportional, reset (integral), and pre-act (derivative). Rules were presented for controller tuning.

Nichols invited Porter to the Taylor Instrument Company in the Spring of 1938, where they discussed the time-lag results. Nichols was curious as to why Hartree did not carry the ideas any further, sensing that he could have made more contributions. But teaching and supervising research of students did not allow Hartree much spare time and in the next chapter we will see that, in addition to the atomic structure research, he also was investigating problems in fluid dynamics.

Other interactions with Industry

As a result of the paper by Hartree and Nuttall, presented to the Institute of Electrical Engineers, industry became aware of the possibilities the differential analyzer provided for their research. When approached, Hartree would try to find a student, often an MSc student, to work on the problem jointly with an industry representative.

The first such problem was concerned with heat generation in dielectrics. For many solid dielectrics the rate of generation of heat in an alternating electric field increases approximately exponentially with temperature. The variation and distribution of temperature was also a subject of considerable importance. Hartree found two MSc students, C Copple and H Tyson, and assigned them to work with Porter on this problem, Porter having received his PhD by that time. When results were obtained, some experimental work was done in the Research Department of the Metropolitan-Vickers Electrical Co., Ltd for confirmation.

Another problem was brought to Hartree's attention by the General Electric Company Limited and the Marconiphone Company. It concerned the distribution of potential in cylindrical thermionic tubes. John Crank and J Ingham were assigned to work on this problem along with Hartree and R W Sloane, of the research staff of the M-O Valve Co., Wembley. The problem was expressed in terms of differential equations of non-dimensional form, and solutions obtained with the use of the differential analyzer for a range of parameters.

Crank went on to complete his MSc in 1938 under Hartree's supervision. His thesis related to the solution of partial differential equations. He obtained his DSc from Manchester University in 1953 but by that time Hartree had left the University. During the war years, Crank was in charge of the newly installed differential analyzer at the Cambridge Mathematical Laboratory then under the control of the government for wartime research. In 1947 he wrote a book, *The Differential Analyzer* [5] which carefully described the construction of the machine, both the regular and small-scale design, and the process of designing a simple set-up. Listed too are the ways in which integrators might be used to obtain many well-known functions such as trigonometric functions or the product of two functions. The references constitute what appears to be a complete list of publications relating to the application of the differential analyzer at that time.

Notes and References

[1] Thaler G J 1974 *Automatic Control: Classical Linear Theory* Benchmark Papers in Electrical Engineering and Computer Science, Series Editor (John B Tomas) (Stroudsburg, Pennsylvania: Dowden, Hutchinson & Ross, Inc)
[2] Hazen H L 1934 *J. Franklin Inst.* **218** 279
[3] Porter A 2000 (private communication)
[4] Zeigler J G and Nichols N B 1942 *Trans. ASME* **64** 759
[5] Crank J 1947 *The Differential Analyzer* (New York and London: Longmans, Green & Co.)

Chapter 10

Laminar boundary layer theory

Hartree was always willing to tackle problems suggested to him where he thought he could make a contribution. One such area was fluid dynamics.

In 1929, when Hartree took on the Professorship in Applied Mathematics, Sydney Goldstein (1903 - 1989) [1], an applied mathematician and a great exponent of aerodynamics, accepted a Lectureship in Mathematics at Manchester along with a Research Fellowship at Cambridge. In 1931 he obtained a regular position at Cambridge, but the two years at Manchester were very influential in his later career in fluid dynamics. During these two years, Hartree was stimulated by Goldstein and *vice versa*. We have already seen in Chapter 6 that Hartree argued for $\beta = 1/3$ in radio-wave propagation whereas Goldstein defended $\beta = 0$.

The equations of motion of a viscous fluid were established in the first half of the 19'th century and are now known as the Navier-Stokes equations [2]. Confusion remained as to the conditions to be satisfied at the wall of a solid boundary: was there slip or no-slip? The latter required a zero velocity at the boundary. In 1904, Prandl recognized that an important question concerned the behavior of the fluid at the wall of a solid boundary. In fact, the region of interest was a thin layer adjacent to the wall and the smaller the viscosity, the thinner was the layer. This small thickness allowed certain approximations to be made to the governing equations.

Prandl's ideas spread slowly, but in 1931 V M Falkner and Miss S W Skan of the National Physical Laboratory, published a paper [3] in which they derived equations and transformed them to a form suitable for application to an airfoil. They were particularly interested in obtaining quantities which could easily be measured experimentally, for cases where the pressure on the boundary was known. In the one-dimensional case of flow passing a solid wall, the equation in dimensionless form, became

$$y''' = -yy'' + \beta(y'^2 - 1), \tag{10.1}$$

with the boundary conditions

$$y = 0, \quad y' = 0 \text{ at } x = 0, \quad y' \to 1 \text{ as } x \to \infty. \tag{10.2}$$

This is a highly non-linear differential equation. Of interest was the value of $y''(0)$ for a solution that would satisfy the boundary condition at infinity, and the behavior of y' for this solution as a function of x. Falkner and Skan investigated the solution in terms of power series, a method not suited to the problem given the need to satisfy an asymptotic boundary condition for large x.

At the suggestion of Falkner, Hartree obtained solutions of this equation on the differential analyzer. Various values of β were considered. Often the values obtained from the differential analyzer were improved using a numerical iterative method which showed that the machine accuracy was about 1 in 1000, or often better. The work had important consequences. Hartree showed that the conditions were insufficient to specify a unique solution for negative values of β. The results were published in 1937 along with a note on a set of solutions of the equation, $y'' + (2/x)y' - y^2 = 0$.

Hartree had just finished some research with J R Womersley on the use of the differential analyzer for partial differential equations in two variables. They showed that, if the derivatives in one variable were replaced by finite differences, resulting in a set differential equations in the other, the differential analyzer could be used *provided* that the boundary conditions satisfied certain requirements. In particular, they showed that the heat conduction problem could be solved in this manner, that a deferred approach to the limit could be used to improve the accuracy. Thus he had an interest in partial differential equations where his method could be used to further research. At this time, Goldstein was editing a book on *Modern developments in fluid mechanics* [4] and encouraged Hartree to consider the boundary layer problem.

The laminar boundary-layer equation is concerned with flow around an object such as a bullet in air or an object in water. The boundary layer surrounds the object, perturbing the surrounding flow, but leaving the main stream at a distance from the object unperturbed. If the object is moving along the x-axis, then the boundary layer equation of an incompressible fluid, in reduced dimensionless variables, is

$$u\frac{\partial u}{\partial x} + v\frac{\partial u}{\partial y} = U\frac{dU}{dx} + \frac{\partial^2 u}{\partial y^2}, \qquad (10.3)$$

the equation of continuity is

$$\frac{\partial u}{\partial x} + v\frac{\partial v}{\partial y} = 0, \qquad (10.4)$$

and the boundary conditions are

$$x = 0, \; v = 0 \text{ at } y = 0, \qquad (10.5)$$
$$u \to U(x) \text{ as } y \to \infty, \qquad (10.6)$$

where $U(x)$ is the main stream velocity. The derivatives parallel to the boundary were replaced by finite differences and integration was carried out along successive normals to the boundary at finite intervals.

Hartree chose a previously investigated problem where $U(x) = 1 - x/8$ for which separation points were known to occur, namely points with reverse flow as shown in Figure 10.1. Since $U(x)$ was given analytically, all the calculations that had been done would use exactly the same values. The aim of the work was twofold: the separation point was to be determined as accurately as possible, as well as the flow through the whole boundary layer so that certain integrals could be evaluated. Because of the need for accuracy, hand calculations to three and four decimal places were carried out. For this Hartree received a grant from the Aeronautical Research Committee to obtain professional assistance from L J Comrie, Director of Scientific Computing Services, and his staff. Even his father participated in some of the early work. Calculations were started up-stream and as the separation point was approached, they required more and more care. The results suggested strongly that a singularity was present at the point of separation.

The second case that Hartree investigated was one where the function $U(x)$ was derived from Schubauer's observed pressure distribution for an elliptical cylinder. Since observations were of limited accuracy, Hartree felt the differential analyzer would be adequate for this problem, but difficulties were encountered and some numerical calculations were needed. The latter clearly showed the reason for the difficulties. The whole behavior of the solution depended on the difference between two comparatively large quantities.

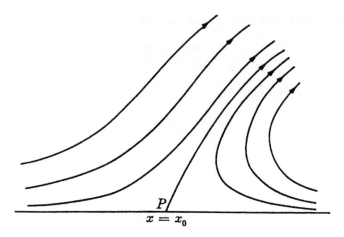

Fig. 10.1 Streamlines at a point of separation. (From *Laminar Boundary Layers* (Edited by L Rosenhead), 1963 (Oxford: Clarendon Press)).

Different investigators had tried to solve this problem but with varying results: some found a point of separation, others did not. In spite of difficulties, Hartree too found no point of separation. At Goldstein's suggestion, he then investigated what modification to the pressure distribution would be needed to yield separation. He found that a comparatively small increase in retarding pressure would lead to separation. A conclusion of the study was that the flow was so sensitive to the pressure distribution assumed that in comparing two methods it was important to make sure that the same distribution was in fact used in each method, otherwise the comparison might not be meaningful.

Two papers were submitted to the Aeronautical Research Committee *Reports and Memoranda* (March 28 and April 4, 1939) and approved for publication, but revisions were interrupted by the outbreak of war and the issue was not distributed until 1949. Often references to this work cite the year 1949, but for present purposes it seems more appropriate to associate it with the year when the work was completed, namely 1939.

Both of Hartree's papers acknowledge indebtedness to Goldstein for his interest and valuable discussions. Goldstein indeed was interested in Hartree's findings, but not much research was done during the war. In 1945 Goldstein was appointed Beyer Professor of Applied Mathematics at Manchester University (the position Hartree had held) and was allowed immediately to build an ambitious Fluid Motion Laboratory, initially in a large hanger at the outskirts of Manchester. As soon as time permitted he followed up on Hartree's observation and investigated mathematically the nature of the singularity that might occur at a point of separation. He concluded that whenever it does occur, the boundary layer equations cease to be valid at and near separation [5], the "importance of which cannot be overstated" according to Lighthill [1].

Notes and References

[1] Lighthill J 1990 *Biographical Memoirs of Fellows the Royal Society* **36** 195
[2] Tani I 1977 *Ann. Rev. Fluid Mech.* **9** 87
[3] Falkner V M and Skan S W 1931 *Phil. Mag. Ser. 7* **12** 865
[4] Goldstein S (Editor) 1938 *Modern developments in fluid dynamics* (Oxford: Oxford University Press)
[5] Goldstein S 1948 *Quart. J. Mech. & Appl. Math.* **1** 43

Chapter 11

Arrangements for war

With war being inevitable, Hartree wrote to Slater on September 1, 1939, the day Germany invaded Poland. His primary concern was the work of his father over the last several years, work that Hartree had not had time to publish. Quite a bit had been done since the last publication, which Hartree now deemed certain he would not have the opportunity to write up. So he thought it would be a good thing if someone outside the country (preferably outside of Europe) had a complete copy, and suggested Slater to his father. He left it to Slater's discretion what should be done with the wave functions and other data.

The results for Na^+ with exchange, needed some explanation in that they did not agree with Fock and Petrashen. He had intended to do independent calculations but had only had time to check some of his father's work and had not found anything suspicious whereas the Fock and Petrashen results were not internally consistent. Hartree had always relied on his father's numerical work entirely. His father could have published them under his name alone but was too modest to agree to this, claiming he did not want to be responsible for the theory, having been an engineer and biophysicist before he turned to atomic structure calculations in his retirement. Hartree claimed he was one of the most reliable computers he had known and that even occasional checks seemed less necessary as he became more familiar with the kind of equations involved. Hartree ended the September 1, 1939 letter with:

> It has been difficult to do any serious work in pure physics lately, with the atmosphere of crisis heavy over everything; and today's news seems to mean the end of such work for me for a time, though I don't know what I'll be doing in its place. For all your country's criticism of our policy in the last 20 years, with much of which I agree, I hope we will

have at least your moral support in the struggle against what seems to us to be the use of armed force as the sole instrument of international policy.

Slater replied promptly to say that he would get someone to put the tables in proper shape, have the Na^+ calculations checked, and publish them under Hartree's signature. He then responded: "We talk about nothing but the war." Hartree's letter had been the first he received from England and it sounded too much like a last will and testament to make him feel comfortable. Of course Slater and his friends were sympathetic and would be happier in a world where England was on top and Germany not too strong, but they did not believe the US should intervene. In this respect he claimed Roosevelt did not represent the country being more in favor of intervention than almost anyone else. He then wrote a rather long assessment of the situation in Europe, ending with the speculation that "By the time you get this letter I suspect Hitler will be having a peace conference for Eastern Europe, writing a substitute for the the treaty of Versailles, ..."

Hartree took a while to reply. He thanked Slater for agreeing to take charge of his father's atomic structure results and for offering to sponsor their publication in *Physical Review*. He then said his father had sent off his unpublished results about a month ago and hoped they had arrived safely (November 18, 1939):

> I had a fear that the censor's office might regard them as some code message and not just what they professed to be and nothing more: if so someone must have had a tough and unremunerative job trying to decode them!

Hartree admitted that in his earlier letter he certainly felt as if he were writing a last will and testament, scientifically speaking. They had been told so often that another war would mean the end of European civilization. And if it were not that bad, it would mean losing touch for years with progress in physics elsewhere. He thought America and Japan might be 3 or 5 years ahead of Britain and developing quite new lines of research. His estimate of getting back to serious research at the end of the war was 1 in 5.

In general, Hartree was glad Slater wrote objectively about his views and not just optimistic comments. He was inclined to agree that one of the real troubles of the last 20 years had been the oscillation between the British and French attitudes towards Germany. He reminded Slater that the French had had their country invaded twice in 50 years, and that the

British, who had not suffered that way, were not really in a position to criticize them for their strong feelings on the matter. But he did not agree with Slater, that the British should take an opportunity, when it arose, to get out of the war if it meant making peace with the present German government. He thought that would simply be asking for future trouble when they might be in a weaker position for dealing with it. He did not think that anyone responsible in the country would take Hitler's word were he to give it.

Regarding the United States, he quite sympathized. He realized there were many important things worth keeping going while Europe was otherwise occupied. Also, he realized that the neutral countries might have a chance to contribute to a settlement that might finally be reached. But what if Hitler were successful and got control of the Navy, making Canada an outpost of the new German empire, where would the US be? This six page hand-written letter is one of Hartree's longer ones.

Slater again replied promptly. It was now December 14, 1939 and William Hartree's results had not arrived. He believed that the sooner Hartree started tracing them the more likely they would be found. According to Slater, English mail had been coming though fairly promptly in less than three weeks. He doubted they had been sunk because there had been no news of liners sunk that would be carrying mail. Maybe Hartree's fears regarding the censor were justified. Slater had contacted the British Consulate who advised Hartree to get in touch with the Censor's Office in London. He went on to say that Hartree's letter had been opened in a curious way – the King's head on the stamp had been neatly cut out leaving a little window for looking into the letter. Slater admitted he too had been feeling pretty low the last time he wrote and that in the meantime his feelings had changed on a number of points. But it is doubtful that Hartree found his letter reassuring. Slater remained pessimistic, stating (December 14, 1939):

> Our skepticism about the war is not about the ends to be accomplished, but about whether they can be accomplished by means of a war, or by any other means we can think of. Let's hope they can.

He ended his letter by describing the scientific progress at MIT – the machine Bush and Caldwell were working on, a new cyclotron that was almost finished, a magnet using enormous currents and huge amounts of cooling that was producing many fine Zeeman effect pictures, to mention a few.

The mystery about the atomic structure results was finally solved. William Hartree had not mailed them to Slater at the same time as when he sent his son a spare copy. Everything was in order except that in January, 1940, William Hartree had a stroke which affected his vision and paralyzed his right arm. In a sense, this was the end of calculations with exchange for the foreseeable future. Hartree did not foresee complete calculations with exchange for atoms much heavier than Cu being done soon, the amount of work becoming excessive. But he thought that there was a good framework of existing data within which to make interpolations. In spite of all this, he expressed the wish (February 1, 1940):

> Someday I would like to see a really heavy atom (Hg, for example) worked out with exchange and Dirac wave functions, but this would probably be a two-year job!

He ended with mentioning that "Miss Swirles," who had been a lecturer at Manchester for a while and part of his staff, was now at Cambridge, and at Hartree's suggestion was looking at intensities of forbidden lines using Dirac wave functions. Hartree had never felt that grafting on the spin-orbit operator to the Schrödinger equation was convincing for qualitative calculations.

In fact, Bertha Swirles had left Manchester in 1938 to return to Cambridge where she took up a Fellowship and lectureship in mathematics at her old college, Girton. There at Cambridge she met Harold Jeffreys and they were to be married on September 6, 1940. Bertha's father died in 1905 and she "was short of male members of the family" so she asked Douglas Hartree to give the bride away in marriage. She was grateful to Douglas and Elaine for making the trip from Manchester to Northampton in wartime conditions. A picture taken in her Mother's garden showing Douglas and Elaine Hartree and Bertha and Harold Jeffreys is shown in Figure 11.1.

With the possibility of nightly air-raids on Manchester, the Hartree children were scattered around the country. Margaret, the oldest of nearly 15 years and already 2 inches taller than her mother, was in a school in the south of England. Oliver, the oldest boy had been at a school in Liverpool which was commandeered and the school then moved to a large country house some 15 miles from Manchester. Their youngest son, Richard, was with his grandparents (Elaine's parents) in Keswick, with the wife and son of one of Hartree's colleagues at the University. Elaine was doing volunteer work in Manchester at a branch of the Citizen's Advice Bureau (CAB) and visiting Keswick occasionally to relieve the mother taking care of her son.

Fig. 11.1 Douglas and Elaine Hartree at Bertha and Harold Jeffreys' wedding in 1940. (Courtesy of Bertha Swirles who in later years was Lady Jeffreys)

In the summer of 1940, after the fall of France and when the invasion threat was real, the staff at the University of Toronto, Canada, offered hospitality to the children of the staff of the University of Manchester. Hartree by then was deeply involved in Ministry of Supply work and knew people who realized how desperate the situation was. The Hartree's decided to take up the offer and sent their three children to Toronto in July, 1940. There the children were separated with Richard being the only one who went to a University family. It was a difficult time for all of them, both as a family and as individuals. There were many sad and poignant memories but, for the children, also opportunities and experiences they would not have had otherwise. The boys arrived back in England in January, 1944 but Margaret, then serving in the Women's Royal Canadian Naval Service (WRCNS), stayed on until the end of the war. The whole experience was hard for the family and, because of their ages, disruptive to Margaret's and Oliver's education. Hartree was very aware of this and that it had been his decision to separate the family. Writing to Bush on August 10, 1945 about the policy of sending children overseas, Hartree said that, for himself, the decision itself had not been difficult though it had been hard to take. He said it was not so much a matter of removing them from risk and personal injury, as from the "sight and sound of destruction and desolation and from the mental shock and atmosphere of suspense and anxiety."

During the period 1941-43, when housing was scarce, the Hartree's made the upper floor of their house available to those they knew – Patrick Blackett, Bernard Lovell, Arthur Porter, Jack Howlett, and Nick Eyres and his wife, Dorothy – who needed somewhere to stay for a while.

Fig. 11.2   Elaine Hartree in her nurse's uniform (1941).

In time Elaine took on work as an auxiliary nurse as well as the Citizens Advice Bureau (CAB). By the time the boys came home, she was no longer working as a nurse but had been put in charge of the CAB organization for the whole of Manchester, a big job. Like many others, Hartree – the Professor and FRS – took his turn at the wartime duty of fire watching for incendiary bombs on a University roof at night.

Hartree himself worked very intensely during the war and was physically drained by it. His son, Richard, recalls how, during a family holiday in the Lake District in 1946 they set off on a climb only to find the exertion was quite a struggle for his father. This was a big change from what he had known before the war. After Hartree's death, Elaine told her son that she felt his wartime commitment had severely weakened his health and shortened his life.

Fig. 11.3 Douglas Hartree (1941).

Chapter 12

Wartime service

During World War II, Hartree supervised two computing groups, neither very large. The one was referred to by Jack Howlett as a "job shop" for differential equations, using the differential analyzer. This work was described in some detail in Hartree's report of 1949 to the Ministry of Supply, entitled *The differential analyzer*. The other group was concerned almost solely with magnetron theory and performance, doing their work on desk calculators. He also chaired the Servo Panel.

12.1 Servo Panel

At a meeting held at Great Westminster House on March 20, 1942 a discussion was held concerning the formation of a Panel on Servo-mechanisms. Present were representatives from divisions of the Admiralty, Ministry of Supply, Ministry of Aircraft Production, and 10 firms including Ferranti Ltd and Metropolitan-Vickers Co., Ltd. The purpose of the panel was to make the different groups aware of servo-mechanical developments in the various units and promote a common terminology. An informal Panel on Servo-mechanisms was formed and at a meeting held on October 30, 1942 Hartree was appointed Chairman of the Servo Panel and Arthur Porter its Secretary. A Servo Nomenclature Panel was set up at a subsequent meeting, also to be chaired by Hartree.

The Servo Panel met almost monthly, much like a colloquium series, with speakers from different firms or government agencies. Attendance varied from an average of 45 to 71 on one occasion. Some 40 organizations were represented at times – Ministries, Services, Companies, Universities, Commonwealth countries and the USA. The first was a talk by Hartree and Porter on *The differential analyzer and its application to servo problems* and given on May 13, 1942. A little over a year later, Hartree chaired the discussion of the first interim report of the Servo Nomenclature Panel, which according to Porter, was the only truly acrimonious meeting of the Panel [1]. In 1945 the Panel had joint meetings with the Manual Tracking panel and, on the death of the Chairman, Hartree took over for a brief time. Chairmanship of these Panels kept Hartree busy and required that he make many trips to London from Manchester by train during the war.

A total of 36 meetings were held, the last on August 24, 1945

12.2 The differential analyzer job shop

At the outbreak of World War II, the differential analyzer at the University of Manchester was the only full-size (8 integrator) differential analyzer in the country. Contact was made with the Ministry of Supply, and arrangements made to have the machine available for any suitable work that might contribute to the national war effort.

The only available operators of the differential analyzer were J Ingham, Hartree's research assistant, and M M Nicolson, a graduate research scholar at the the University. For a time they were working largely on their own since Hartree had joined the Projectile Development Establishment (PDE), an anti-aircraft rocket design center then at Fort Halstead. With only two in the group the application of the machine was severely restricted, but in the early years, knowledge about the capabilities of the machine among those in the country concerned with technical problems was restricted primarily to ballistic problems. Thus, for the problems that arose during the first year of the war, a group of two was adequate.

In the summer of 1940, Nicolson was called up, his reservation for work on the differential analyzer not being accepted. The loss of his services was severe, coming at a time when the possibilities of saving many manhours were being recognized. Hartree found a successor, namely Nicholas R Eyres, related to Elaine Hartree, who had taken a Mathematics degree at Cambridge and was a school teacher. Miss Phyllis Lockett (later Mrs Nicolson) also joined the group. Early in 1941 Hartree was transferred to the Department of Scientific Research of the Ministry of Supply, and became freer to supervise and take part in the work of the group. Ingham left and to replace him, Hartree asked his former student, Jack Howlett, to join the team since he had acquired some experience with the differential analyzer while working for the LMS Railway. Later in 1941, the unit was

Fig. 12.1 The SR(A) group with the differential analyzer. At left, from front to back: Jack Howlett, Nicholas R Eyres, J G L Michel; center, Douglas R Hartree; right, Phyllis Lockett Nicolson.

further strengthened by the appointment of J G L Michel. In 1942 the group was reorganized so as to come directly under the Scientific Research Headquarters of the Ministry of Supply and was named SR(A) for brevity. A picture of the group is shown in Figure 12.1.

The problems brought to the group at Manchester were by no means straight forward, either mathematically or as an application of the differential analyzer. The group would discuss the difficulties with Hartree. More often than not, he would say he needed to think about it and would turn up the next morning with the method of solution together with several pages of computation to check that it would work. The story is told [2] of an occasion when he was asked if the differential analyzer could help with a problem concerning magnetic mines. Hartree's reply was "Go and have breakfast while I think about the problem." By the time the colleague returned, Hartree had the answer – the machine was not needed since he had already come up with the solution.

Howlett described Hartree as a collaborative and inspiring leader. But his work demonstrates that he also had exceptional abilities to see how a problem could be expressed in mathematical terms.

The work performed by the SR(A) group can be classified into two categories: ordinary differential equations, often non-linear and/or with two-point boundary conditions, and partial differential equations. To indicate the tremendous range of problems that were investigated, some of the more important ones are mentioned here.

Ordinary differential equations

I. Ballistics

In the early days of the war, the differential analyzer work was concerned with ballistics of various types.

(1) *External Ballistics* in which a projectile was considered as a particle and a trajectory was required. Already prior to the war, Hartree and Porter had done exploratory studies. They had come to the conclusion, that through special precautions, a consistent accuracy of 1 in 500 could be attained, but that it would be difficult to guarantee higher accuracy. With the outbreak of the war, the problem was revisited given its urgency. The work was limited to a standard plane trajectory, with no wind and standard density variation with height, but no real progress was made.

(2) *External Ballistics with weighting factors.* Now the trajectory calculations involved a resistance function and the variation of the derivative of the resistance function through the velocity of sound. The differential analyzer was much better suited for this problem with the attainable accuracy more than sufficient.

(3) *Projectile as a rigid body.* The only problem in this category was the motion of a rotating rocket fired from an aircraft. This problem differed from those encountered with guns in that the velocity of the rocket relative to the aircraft was considerably less than that of the aircraft. Thus the path of the rocket after launch depended on the aircraft velocity and attitude. In certain circumstances it would leave the projector with a yaw and its motion, at least initially, might be closer to the line of sight than the correct direction. It was desirable to know whether by choosing aerodynamic characteristics of the rate of spin, the rocket would, effectively, proceed toward the target independent of the aircraft attitude.

(4) *Internal ballistics* was concerned with the burning of a propellant in a gun and with the motion of the projectile in the gun barrel during the burning of the propellant.

Most of the work on ballistic problems early in the war was done by Mr J Ingham and Mr M M Nicolson.

II. Automatic following of targets

Through various contacts and lectures on the differential analyzer by Hartree to the newly formed Servo Panel of the Ministry of Supply that he chaired, the possibility of applying the machine to the study of the performance of control systems and of obtaining guidance on their general design became known to those concerned with automatic following, fire control, and similar purposes. This led to a considerable amount of work, an example being the automatic following of targets.

Radar equipment was supposed to give a continuous record of the position of a target, following an unknown course, without any manual intervention. This information was to be used to move, for example, a gun or searchlight in a way that depended on the position of the target. Generally, a signal was sent from a transmitting aerial, reflected at the target, and from the received signal strength, information about the position of the target was deduced. An important point needed to be considered. If the equipment was to control gun-fire, the guns would have to follow the *future* positions of the target. The latter required predictions from current information and hence that the information be smooth.

A phenomenon of considerable concern, was "high speed fading" of signals. In their work, signals were pulses of the order of a microsecond duration at intervals of about 400 microseconds. In the absence of fading, the variation of signal strength received described the motion of the target and on a long time scale, could be considered continuous. But with fading, the signals first needed to be smoothed before they could be sent to an automatic following apparatus. A variety of smoothing and predictor methods were considered.

Several requests were made by Arthur Porter at the Ministry. An important example was work done on a proposed control system for G L Mark III, an anti-aircraft artillery control radar set. The work on the differential analyzer showed first that, with perfect radar signals, substantially improved performance could be achieved by modifying some design parameters, secondly that the effect of fading signals would negate the performance of the entire system, and thirdly that any attempt to smooth the fading systems would degrade the performance of following a target with perfect signals. As a result, the development was never undertaken, thus saving an immense amount of effort. Porter considered this a significant contribution.

III. Radio Propagation

The propagation of radio waves in the lower atmosphere had been a subject of considerable interest with the recognition that the vertical gradient of the refractive index for short radio waves could be negative and large enough to make the curvature of the ray greater than that of the earth's surface. As a result, under suitable meteorological conditions, radio ranges could be many times the range of a geometric horizon. Propagation under such conditions was sometimes called "anomalous propagation" though, according to Hartree, there was really nothing anomalous about it.

Work was undertaken to investigate the relation between meteorologic conditions and radio wave propagation. Hartree was familiar with this topic, having published three papers on the subject during the period 1923-1931. His early work had shown that there were two ways of approaching the problem of wave propagation in a stratified medium: a ray treatment or a wave treatment. The SR(A) group did work undertaking the ray treatment whereas a group at Cambridge applied the wave treatment. A report on *Practical methods for the solution of the equations of tropospheric refraction* by D R Hartree, J G L Michel, and Phyllis Nicolson, was presented to the Physical and Meteorological Societies' Report on Meteorological factors in Radio Propagation, in 1947.

IV. Under water explosions

The Admiralty Mine Design Department requested an investigation of the behavior of a bubble of gas formed by the detonation of an explosive charge in a large body of water. The bubble was assumed to be spherical at the moment of formation. The buoyancy and high pressure of the enclosed gas would cause the bubble to rise and expand. The force-system acting on the bubble was such that it departed from being spherical. What was wanted was a detailed account of the motion and stability of a non-spherical submarine bubble.

V. Motion and Stability of an aircraft

A very large amount of work was done by the differential analyzer on the motion of an aircraft resulting from the application of some disturbance on the action of the controls, automatic or otherwise, in attempting to correct the effect of the disturbance. Examples of disturbances were the sudden failure of one engine or the movement of a control surface such as a rudder or aileron. The problem was well suited to the machine, involving sets of differential equations, not necessarily linear, with four or more variables. Over a period of two years, about 1,500 runs were completed, requiring about six months of work implying that about 10 runs could be made in a day. Previous experience had shown that numerical integration of a typical set required at least 4 days. The problem was proposed by the Aerodynamics Section of the Royal Aircraft establishment, Farnborough.

Partial differential equations

The differential analyzer was a machine for integrating *ordinary* differential equations, whereas problems dealing with space and time often led to *partial* differential equations. Hartree had reported to the International Congress of Mathematicians, Oslo, in 1936 on this application and evaluated its use together with Womersley in 1937. Now he had an opportunity to apply his ideas to a wide range of problems.

I. Integration methods for heat flow problems

The simplest heat conduction problem is that of a uniform slab or bar of given initial temperature whose ends are maintained at some given temperature. If the distance along the bar is represented by the variable x and the temperature of the bar at time t and position x by $\theta(t, x)$, then, assuming the composition of the bar does not have conduction properties that depend on the temperature, the temperature of the bar is given by the partial differential equation (said to be of parabolic type) that can be reduced to

$$\frac{\partial \theta}{\partial t} = \frac{\partial^2 \theta}{\partial x^2}, \qquad (12.1)$$

with boundary conditions

$$\begin{aligned} \theta &= \phi(x) \quad \text{for} \quad t=0, 0 \leq x \leq 1, \\ \theta &= f_0(t) \quad \text{for} \quad t>0, x=0, \\ \theta &= f_1(t) \quad \text{for} \quad t>0, x=1, \end{aligned} \qquad (12.2)$$

where $f_0(t)$ and $f_1(t)$ are given functions of t and $\phi(x)$ is the initial temperature. The bar is assumed to be of unit length.

To solve this equation on the differential analyzer, the time derivative could be replaced by a finite difference approximation. Hartree was a firm believer in centered differences. In this case, the difference, $[\theta(t + \delta t, x) - \theta(t, x)]/\delta t$, was taken as an approximation to the $\partial \theta(t, x)/\partial t$ at $t + \delta t/2$. The partial differential equation then required $\partial^2 \theta(t, x)/\partial x^2$ also at $t+\delta t/2$. This he approximated as the average of $\partial^2 \theta(t, x)/\partial x^2$ at t and $t + \delta t$. The result was a formula that led to a differential equation at $t + \delta t$ rather than an explicit expression, but with his differential analyzer, this was not a problem. Since the solution was known at $t = 0$, repeated application of his formula made it possible to produce a solution marching forward in time as shown in Figure 12.2. Hartree referred to this as Method I.

Another method, called Method II, was obtained by replacing the second order space derivative by a finite difference. This results in a system of differential equations as shown in Figure 12.3. Since θ_0 and θ_N are known, there are $N-1$ unknown functions of t. These are $N-1$ simultaneous equations that need to be integrated together. Since an integrator was needed for each first-order differential equation, no more than 8 functions could be accommodated but when the variation with x was smooth, only three functions were used and the method was referred to as a "four-step"

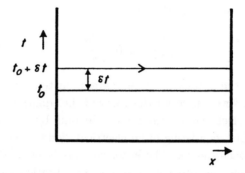

Fig. 12.2 A graphical depiction of a solution of a two-dimensional partial differential equation by Method I. (From Hartree D R 1949 *Ministry of Supply, Permanent Records of Research and Development Monograph* **17.502**)

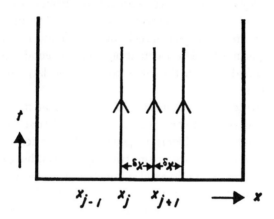

Fig. 12.3 A graphical depiction of a solution of a two-dimensional partial differential equation by Method II. (From Hartree D R 1949 *Ministry of Supply, Permanent Records of Research and Development Monograph* **17.502**)

method. The latter method was applied to a number of heat flow and diffusion problems.

Almost every process in the iron and steel industry was concerned with either heating or cooling of large masses of metal. In 1939, the National Physical Laboratory had reported, upon request of the Thermal Treatment Subcommittee, on the properties of a number of steels. When the information became available, it became clear that conventional theoretical methods were incapable of solving the diffusion equation, since the thermal properties of steel depended on the temperature. The simplest, one-dimensional case, led to the equation

$$\frac{\partial \theta}{\partial t} = D(\theta)\frac{\partial^2 \theta}{\partial x^2}, \qquad (12.3)$$

where $D(\theta)$ was the diffusivity of the material at temperature θ.

Two machines were used for the study of the equation – the differential analyzer and an electrical computer, specially built by Metropolitan-Vickers Electrical Co., Ltd, consisting of an electrical "ladder network" of a series of resistance capacity loops. The latter could only be used in cases where $D(\theta)$ was constant, but provided a means of calibration for the differential analyzer. Using Hartree's experience in the use of time-lag in control systems, a rapid practical method for obtaining the temperature at the center of mass from a time-lag solution was developed. At the same time, experiments were performed measuring temperatures inside large steel masses under conditions of industrial heating and cooling. The first paper, *Variable heat flow in steel*, focussed largely on the comparison of theoretical results with experiment. The calculations showed good agreement with experiment obtained both in the laboratory and on large-scale masses in industrial furnaces.

The second paper, *The calculation of variable heat flow in solids*, was both more general and more theoretical. It stressed that the method was appropriate for heat flow problems, in general, and practicable for a number of geometries, such as a square prism, two-dimensional heat conduction in a cylinder without restriction to axial symmetry, as well as one-dimensional heat flow in a slab or radial heat flow in a cylinder. The one-dimensional case with constant diffusivity and special boundary conditions which lead to analytic solutions was used as a means of determining the errors of the approximate method in which the second order space derivative was replaced by a finite difference. In addition to a number of different geometries, the paper considered different boundary conditions, such as the application of Newton's cooling law where the surface heat flux is proportional to the difference in temperature at the surface and the ambient temperature of the surrounding medium. Special attention was also given to heat flow while the steel was going through a transformation (such as liquid to solid).

Results of calculations were compared with experiment for quite different situations, which included:

(1) The heating of a 14-ton octagonal ingot of mean diameter of 28 inches, treated as a radial problem. For its solution the radius was divided into

6 equal intervals.

(2) The heating of 1.75 and 3.5 inch bars for which experimental measurements could be made under closely controlled laboratory situations.

But the most ambitious undertaking was the cooling of an ingot in a mould, including radiative heat transfer. In the casting of an ingot, molten steel was poured into a mould. Soon after the outer surface of the steel became a solid, a gap formed between the outside surface of the ingot and the inner surface of the mould, increasing in width as the mould heated and expanded and the ingot cooled and contracted. From the time when the gap was formed, transfer of heat from the cooling ingot took place by conductance through the ingot and the mould walls, and by radiation across the gap. The results of this research suggested ways in which the cooling process could be achieved more efficiently.

II. Diffusion and shock wave equations for Tube Alloys Directorate

At the suggestion of Rudolf E Peierls, then at the University of Birmingham, a request from the Directorate of Tube Alloys (British bomb project) was made of a study of a non-linear diffusion equation. The equation was too complicated for the second method (the analyzer capacity being insufficient) and the first method needed to be used, but an analytic starting solution was required along with numerical integrations. This was one of the most elaborate investigations undertaken by the SR(A) group. The problem was related to the production of the critical mass of uranium for sustaining a chain reaction.

Sometime in February or March of 1940, knowing that bombardment with slow neutrons of the ^{235}U isotope of uraniam would result in fission that could set off a chain reaction, Robert Frisch and Peierls set out to estimate the critical mass for an explosion [3]. They came up with an estimate of about one pound. This assumed a certain fission cross-section that needed to be measured for a more reliable estimate. Since naturally occurring uranium ore contains only about 0.7% of this isotope, it needed to be separated from the other ^{238}U isotope. This could be achieved through gaseous diffusion in an isotope plant. A problem which gave them difficulty was estimating the equilibrium time that would be needed. This is the time it takes from starting up to producing output of the designated concentration. Peierls wrote to Hartree on November 4th, 1941 [4], mentioning that he recently came up against a partial differential equation that he thought Hartree's differential analyzer could solve though it was non-linear. He said that, of course it was associated with war research and, if necessary, he could arrange for an official first-class priority. Hartree responded a few days later that the differential analyzer was fairly busy, but that there were gaps, and they could fit the project into one of the gaps. Or, if the Ministry of Supply gave Peierls' work priority, they would postpone other work.

From this beginning the work grew. Over the next four months, Peierls and Hartree exchanged extensive correspondence about the approach and method. In March 1942, results were exchanged between Klaus Fuchs (who was working for Peierls), Eyres and Howlett. Contacts continued by correspondence, telephone, and visits.

On September 14, 1942, Peierls wrote to Hartree about another problem he might wish to tackle while awaiting data. It dealt with the long-range end of a blast wave produced by a TNT explosion. It was related to Peierls' work on the mathematics of actual bomb explosions. The calculations were primarily about the propagation of blast and shock waves and the efficiency of the explosion. These were Howlett's calculations on explosions. On January 29, 1943 Howlett sent Peierls the Manchester group's report. On April 21, 1943, Peierls, in turn, sent Hartree the final report for the Directorate of the Tube Alloys, and thanked him for the group's contribution, acknowledging that the work would not have progressed as far without it.

In all this correspondence, the applications were not mentioned. Reports and communications were nearly all mathematical and secrecy was maintained.

Notable, in spite of the war, was that the Royal Mail was giving next day service between Birmingham and Manchester.

General comments

In his 1949 report to the Ministry of Supply, Hartree made three general comments on the collective experience of the SR(A) group.

(1) *Value of the work.* If the equations were simple enough to be handled by formal means, nothing more was needed. But formal solutions were not always possible, nor were they always easy to interpret. If the equations were within the capacity of the machine, it was usually found that the machine provided results with the expenditure of only a small fraction of the labour required to build a system.

(2) *The atmosphere of freedom.* Hartree clearly believed that a small group, working together, unhindered by red tape and free of admin-

istrative matters such as monthly reports, greatly contributed to the efficiency of the group.
(3) *The capacity of the differential analyzer.* More than once during the war, a problem was submitted that was beyond the capacity of the machine (8 integrators). To Hartree at the time, it seemed desirable, that if another machine of this type was to be installed in the country, it should be substantially larger, having at least 12 integrators and preferably 16. He mentioned the new machine at MIT with 18 integrators, with a possibility of expansion to 30.

Not mentioned in this report were solutions to other partial differential equations that had been solved on desk calculators. One such problem was the fire (burning) problem of great importance in wartime conditions. This was another heat-conduction problem, very non-linear and with a temperature-dependent term. Hartree was asked to look into it [5]. He could see a general way of how the problem might be treated but could also see that a lot of detailed work would be needed. Looking for the best people, he had no hesitation in suggesting John Crank, a former student of his, and Phyllis (Lockett) Nicolson.

This led to the development of Method III, a combination of Methods I and II now known as the famous Crank-Nicolson formula for parabolic equations [6]. When numerical methods are used (as distinct from differential analyzers), both the time *and* the space derivatives need to be replaced by finite differences. A computationally simple, explicit method had been suggested by others but gave rise to an oscillating error. In this paper Crank and Nicolson present an implicit method requiring an iterative process at each step. As a result, more computations were needed at each step, but the oscillations were removed and much bigger steps in time could be used. Hartree was intimately involved in the work. The authors acknowledge his many helpful suggestions.

The effect of secrecy clearly affected his research on partial differential equations during wartime. The first paper on the *Variable heat flow in steel* seems not to have been covered by the Secrecy Act, and was published in 1944. In 1943, Hartree was asked to give the XXX'th Kelvin address to the Institute of Electrical Engineers. He talked about the application of the differential analyzer to electrical engineering problems. None of the wartime problems were included as examples, though there was a brief, general description of how the differential analyzer could be used to solve heat conduction problems.

12.3 Magnetron research group

Hartree supervised another group consisting of Phyllis Lockett (also in SR(A)), David Copley (a schoolmaster from Sheffield) and Oscar Buneman. When World War II started, Oscar was Hartree's student, but as a foreign national he was interned in Canada. By means of a letter from Hartree, no doubt mentioning his considerable scientific abilities, he was released after about nine months. Upon completing his PhD in 1940 he continued to work with Hartree as a research assistant in the magnetron group until 1943. The work of this group was done for the Committee on the Co-ordination of Valve Development (CVD), assisting with the development of radar.

Today it is almost impossible to think of a world without radar which led to the development of microwave ovens, now part of our everyday lives. Radar, or radio detection and ranging, played a crucial part in the defense of Britain. Ground based radar was used extensively to give early warning of enemy aircraft, and airborne radar for the detection of ships and the interception of aircraft, particularly at night.

In the summer of 1939, Edward G Bowen, who had been a student of Appleton's and played an influential role in the development of radar since 1935, was asked by the Chairman of the Committee of Valve Development (CVD), what wavelength was most appropriate for airborne radar [7]. He did a back of the envelope calculation and came up with 10 centimeters. It was not long before CVD contracts were placed with Professor Oliphant at Birmingham University and others for the development of transmitting and receiving tubes for 10 centimeter wavelengths. Oliphant had studied with Rutherford at Cambridge University and his staff were mostly nuclear physicists, not electronic engineers, which he argued was an advantage.

Two members of his staff were Henry Albert Howard Boot (1917-1983) and John Turton Randall (1905-1984). At first they attempted to improve on existing magnetrons, but then, in a flash of inspiration saw that the way to go was with a number of resonator cavities arranged in a circle with slots connecting them to the central interactive region surrounding the cathode. They built the device, with the first production models becoming available in July 1940, with outstanding performance. But the underlying theory was somewhat of a mystery and many questions remained unanswered such as resonator size, number of resonators, operating conditions, and output circuit. Improvements continued. An important one was suggested by J Sayers, also at Birmingham. He proposed that certain modes of oscillation could be eliminated through the use of "straps." When straps were applied, the increase in output power was striking. After the war, Boot and Randall published a report on their investigations [8].

In the early months of 1940, the situation in Britain was desperate. The air attacks had not yet come but no one was optimistic. At that point it was suggested that Britain should share its secrets with the USA and Canada in return for technical help and production. This was a controversial recommendation but A V Hill (leader of the "Brigands" in WW I (Chapter 2)) was sent to Washington to make discrete inquiries. He sent back encouraging reports but recommended that there be full unconditional disclosure, being confident the USA would react whole heartedly. On August 29, 1940 a mission left for Washington, including Bowen whose book *Radar Days* [7] gives an interesting account of the mission. Among the American scientists involved in the decision making on how to participate in the development were Vannevar Bush, now President of the Carnegie Institute and ex-Dean of MIT, and Karl Compton the President of MIT. Soon MIT was selected as the institute in which a new laboratory was to be created, with physicists and engineers to be recruited from across the country. It became known as the MIT Radiation Laboratory, funded by the National Defense Research Council (NDRC) later renamed the Office of Scientific Research and Development (OSRD).

According to Slater [9], war came to the MIT physics department in October, 1940. A nuclear physics conference was scheduled which was well attended, but many assembled behind closed doors and talked about the plans for a new microwave laboratory. The physics department was not consulted. This was part of Compton's plan for the Institute to host the laboratory but make no effort to influence policy or science. However, Slater was soon contacted to help in the understanding of the theory of the magnetron.

Thus Hartree and Slater were again working in the same area. In later years, when Dr Robert L Kyhl, one of the early graduate student recruits, was asked in an Oral History interview for the Institute of Electrical and Electronics Engineers [10], how significant was the work of John Slater to the ultimate development of magnetron designs and to the comprehensive theory of magnetrons, he replied "Eventually profound," but went on to say that Hartree "is the father of magnetron theory." The extent of his contributions are not well known to scientists in general, although they are mentioned in the more advanced books on resonant cavity magnetrons [11] and in publications of formerly classified work [12], part of the the MIT

Radiation Laboratory Series.

The cylindrical magnetron was studied by Hull in 1921 as an interesting laboratory device. It consisted of a cylindrical anode surrounding a concentric cathode with a voltage V between them and a magnetic field parallel to the axis of the cylinder of strength B. In the static case (where the potential is only a function of r and not of time), the motion of electrons can be analyzed (see [13; 14]). In the case where the radius of the cathode is small compared to the radius of the anode, the problem simplies further. For a given voltage, there is a critical magnetic field strength. When the magnetic field is absent (or zero) the electrons proceed directly from the cathode to the anode, for a small field the orbit is an arc striking the anode, but for large fields the orbits are circular with the electrons returning to the cathode without having reached the anode. In the latter case, no current can flow. In between there is the situation where the orbit just touches the anode and this is the critical field strength. This critical relationship between voltage V and field strength B is referred to as the "Hull cut-off condition" [13]. It can be expressed through an equation in which the voltage is a parabola with respect to B. This derivation omits some practical considerations. In actual fact, many electrons can be emitted from the cathode and may form an electron cloud most dense near the cathode. The charge density of electrons within the cavity is known as *space charge*. Under steady state conditions, electrons leaving the cathode acquire an angular velocity, and when the magnetic field is high enough, the electron cloud rotates about the cathode almost as an angular body.

One of the first questions Hartree's magnetron group considered was whether the circulating current due to the motion of the electrons might not itself produce a magnetic field comparable to the external magnetic field. Because of the radial symmetry, the problem could be set up in terms of ordinary differential equations that were solved by the numerical methods Hartree had published in 1933. They found that the magnetic field produced was of the order of 3% and hence not important. The study, including many pages of handwritten equations, was reported in CVD Magnetron Report No 1, the first of many reports that Hartree wrote.

Though the static, cylindrical magnetron was amenable to theoretical analysis, it was not the device that needed to be investigated. The arrangement that Boot and Randall first considered was a 6 cavity magnetron, though they also tested an 8 cavity magnetron. The whole anode block was machined out of a solid block of copper which gave it good electrical and thermal properties. The first cavity magnetron had a tungsten filament as cathode but later versions used oxide-coated cathodes.

The cavity magnetron did not lend itself to analytic analysis. Extensive numerical calculations were performed by Hartree's magnetron group. They used a self-consistent field (SCF) method similar to the one he had used for atomic structure calculations, but computationally much more difficult:

(1) Assume a likely static electric field distribution
(2) Calculate electron orbits in this field.
(3) Calculate the space charge field produced by these electrons
(4) Compare the field calculated in (3) with that assumed in (1)
(5) Repeat (ii) using the field found in (3)

The above was repeated until the computed electric field agreed with the assumed.

Under oscillating conditions, the anode voltage is like a travelling wave of the form $V_0 + V_1 \cos(n\phi - \omega_0 t)$. In a rotating field approximation, the potential was considered to be a function of the polar co-ordinates (r, ϕ) relative to axes that were themselves rotating with the field at the anode. The differential equations for the orbits included a second-order differential equation with two initial conditions, and two first-order differential equations where for the last equation, the intial condition was the angle at which the electron was ejected from the cathode. For each field distribution, a number of orbits leaving the cathode at various angles were computed. The calculation of orbits was work that could have been done on a differential analyzer since the numerical integration was rather laborious and not rapid. Several attempts were made to design satisfactory machine set-ups to handle these equations on an eight-integrator machine, but without success. The best set-up from the point of machine running required 12 integrators. A differential analyzer would have been even more valuable for answering the question whether the electronics would be much affected by other components of the field rotating at other angular velocities. This complicated the problem to the point where 16 integrators or more would have been needed. So the analyzer could not be used and, instead, Hartree set up his group of three "CPU's" (as Buneman called them [15]) to work on mechanical desk calculators.

For a method of solution, he decided on what is now a classical particle simulation. The orbits for a large number of particles in a field were integrated. The field could be revised in accordance with the instantaneous charge density at each step, or only occasionally.

Both one and two-dimensional simulations were performed using three

Marchant mechanical add-and-shift machines. Hartree noted that the problem was amenable to parallel computing. Several hundred orbits would be shared out among the three of them. According to Buneman [15], Phyllis Lockett always was the fastest. Even so, they were a billion times slower then modern CPU's in 1990.

Hartree provided an elegant solution to the difficulty that arose when time-centered differencing was used for the Lorentz equation of motion for an electron in a magnetic field, making the problem implicit. Hartree set out instructions to the three "CPU's" in the form of a program with go-to's and loops. These were not called "looping" in those days. In fact, Phyllis referred to them as "knitting."

A major problem was solving Poisson's equation in two dimensions for the potential from the charge density. Hartree introduced them to Southwell's relaxation technique, an iterative method where, for each point in the two-dimensional region, current estimates were improved until changes were sufficiently small. He provided them with large plastic sheets on which they could record the two-dimensional potential array and on which they could easily rub out work to record their improved guesses. They found it far from relaxing! In fact, as Buneman stated, "It was frustrating to chase residuals to the boundaries." As a result, iterative methods were abandoned and Hartree switched to the direct Fourier method. It turned out that a modest number of harmonics were sufficient. Eventually, plausible particle-field configurations emerged, showing the four-spoke wheel that rotates in the magnetron as shown in Figure 12.4.

In fact, magnetrons can oscillate in different "modes." In Figure 12.4 the phase difference between adjacent segments is π and so this is called "π-mode." But other modes are also possible and were investigated leading to what are known as "Hartree harmonics." The strapping mentioned earlier was a means of preventing mode changes.

By the self-consistent (SCF) method, Hartree and his group found that, under oscillating conditions, electrons tended to divide into two groups, useful and non-useful. The latter were accelerated by the DC field and rapidly returned to the cathode. The former were retarded by the magnetic field, thereby releasing energy to the RF field.

In the two dimensional simulations, Buneman found it beneficial to be a human CPU. A computer would grind out millions of zeros without objecting or giving some warning, but as a human he observed that his particles were shunning certain regions in the field. This led to the discovery of the threshold criterion for magnetron operation. This was analyzed

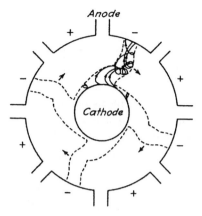

Fig. 12.4 Electron orbits in an oscillating magnetron. Dotted lines represent boundaries of the space-charge cloud, rotating synchronously with the electromagnetic wave on the anode. The solid curves show electron paths within the rotating wheel. The magnetic field is perpendicular to the plane of the page. (CVD Magnetron Report No 41 and Reich H et al 1953 Microwave Theory and Techniques (New York: van Nostrand)).

further by Hartree (and others) and a formula derived. This condition, sometimes referred to as the "Buneman-Hartree criterion"[16] but, according to Slater [14], generally known as the "Hartree condition," corresponds to the voltage below which, oscillations cannot be maintained. Sometimes it is also referred to as the "Hartree equation"[17]. It is a general definition for all types of magnetrons, but for the cavity magnetron it is the voltage at which the angular velocity of the outer electrons of the "wheel" coincides with the angular velocity of what Hartree (in CVD Magnetron Report No 10) referred to as the "stationary wave of oscillating potential on the anode." The condition depends on the mode of oscillation.

In the absence of oscillations, the space charge is circular in shape, rotating about the cathode. Its shape must change for oscillations to start. When the speed of rotation of the space cloud is approximately synchronous with the resonant frequency of oscillations on the anode, it exhibits instability and tends to change to a spoked wheel. Hartree calculated the minimum voltage for this to happen for a particular magnetron. It is called the "instability voltage" but no general formula was derived.

Hartree and his group investigated many aspects of magnetron operation such as the efficiency and transient behavior. Oscar Buneman himself studied the "small amplitude" theory of the magnetron, for which he used the differential analyzer for obtaining 50 different solutions, and wrote his own CVD report [18].

When the problem of the magnetron was presented to Slater, he states it was immediately obvious to him how to proceed because of the resemblance to the self-consistent field in an atom [9]. He resolved to undertake such a calculation in order to see if the result would include not only the time-independent space charge, but also an oscillating space charge which would correspond to the radio-frequency output of the magnetron. This was quite a calculation, but after a few weeks he had a satisfactory answer. Furthermore, the calculations explained the high efficiency of the magnetron.

It was reported back to Slater that Hartree too was working on the problem, again using the self-consistent field method, coming to the same conclusions. Yet his results differed by a factor of 2 somewhere. He knew Hartree well enough to know he would not have made a mistake, but was mystified why his results were wrong, since they had been checked by a colleague. It turned out that both had made a mistake by introducing a factor of 2, each in a different place, so the final results agreed but both were wrong.

Many members of the Radiation Laboratory were treating the microwaves by methods known as "equivalent circuits," based on the analogy between oscillating cavities and ordinary electric circuits. Once the general principals of the magnetron operations were clear, Slater felt it was necessary to go further. He had treated a microwave cavity as a real cavity, dealing with solutions of Maxwell's equation, and it was clear to him, the same needed to be done for the rest of the microwave circuit. He felt these methods should be made available to the members of the laboratory, and took time off to write the book *Microwave Transmission* [19] which could be published as an unclassied book.

Hartree never published any of his findings on magnetron theory in research journals. During the war, he wrote numerous, highly technical CVD magnetron reports which were secret. By May, 1945 just at the end of the war, Hartree had already moved on to become involved in the development of computers.

Notes and References

[1] Porter A 1971 *IEE Centenary Lectures 1871 – 1971 Electrical Science and engineering in the service of man*, pp 136

[2] Marsh P 1978 *New Scientist* **80**, No 1134
[3] Peierls R 1985 *Bird of Passage* (Princeton, NJ : Princeton University Press)
[4] Communications between Hartree and Rudolf Peierls, part of the Tube Alloy Project, Public Records Office, London file AB 1/574, R Peierls, personal files, HARTREE.
[5] Howlett J 1996 *Advances in Computational Mathematics* **6**.
This issue on *Partial Differential equations and applications* was dedicated to John Crank's 80'th birthday. It includes a *Foreward* by Jack Howlett.
[6] Crank J and Nicolson P 1947 *Proc. Camb. Phil. Soc.* **43** 50
[7] Bowen E G 1987 *Radar Days* (Bristol, England: Adam Hilger)
[8] Boot H A H and Randall J T 1946 *J. IEE* **93** 928
[9] Slater J C 1975 *Solid-State and Molecular Theory: A Scientific Biography* (New York: John Wiley & Sons)
[10] Search for "Kyhl" at http://www.ieee.org
[11] Boulding R S H 1952 *The Resonant Cavity Magnetron* (London: George Newnes Limited)
[12] Collins G B 1948 *Microwave Magnetrons* (New York: McGraw Hill)
[13] Hull A W 1921 *Phys. Rev.* **18** 31
[14] Slater J C 1950 *Microwave Electronics* (New York: van Nostrand Co., Inc)
[15] Buneman O 1990 *A History of Scientific Computing* Edited by S G Nash (New York: ACM Press) pp 57
[16] Buneman R, Barker R J, Peratt A L, Brect S H, Langdon A B, and Lewis H R 1994 *IEEE Transactions on Plasma Science* **22** 22.
[17] AccessScience, McGraw-Hill, http://www.accessscience.com, search for "Hartree"
[18] Buneman O 1944 CVD Magnetron Report No 37
[19] Slater J C 1942 *Microwave Transmission* (New York: McGraw Hill)

Chapter 13

Dawn of the computer era

Arthur Porter spent 1937-1939 with Bush's group at MIT. He informed Hartree of the design of a new machine being built by Bush and Caldwell using electronic circuits. In his reply [1], Hartree expressed great interest in what Porter had to say. He mentioned that Blackett had returned from America with a story of 2,000 vacuum tubes (valves) in the new MIT differential analyzer; a letter from Caldwell some time earlier mentioned only 1,200 tubes and Hartree wondered whether the plans for the machine had grown in the meantime. He said that, in his own lab, Blackett and F Williams had designed an automatic curve follower for the input table that was electronic [2]. It was a photo-electric device that followed the input curve in a succession of small steps. This had proven to be unsatisfactory for driving the integrating disks on the machine. To overcome this, they detected the slope of the input curve by using a single photocell illuminating two points on the curve alternately. Hartree claimed he did not pretend to understand the electrical side. He mentioned that they had only the experimental version, that it was a bit temperamental and needed keeping an eye on, but otherwise was satisfactory and worked quite well. But his biggest curiosity was in another machine:

> Do you happen to have gathered what the US Ordnance Department machine at Aberdeen (I think) is used for? Blackett said that he had heard that it calculates trajectories in only about 3 times as long as the shell takes to traverse them, which I find difficult to believe. I am interested in what it is used for, in case, in event of war, there may be work of a serious nature for our machine; I would prefer to keep away from such work in the ordinary course of things, but the experience of the last month[1] has shown that it may be as well to prepare for such applications.

---

[1] Chamberlain's meeting at Munich with Hitler

This machine later was known as "ENIAC."

As in the US, the British development of electronic computers was strongly influenced by the wartime activities [3]. The British Post Office Research Laboratories developed Colossus, the first programmable electronic computer, in complete secrecy, to help break top-level German machine ciphers. Design started in March 1943, under the direction of M H A Newman (1897-1984). The Mark I Colossus, with 1,500 vacuum tubes was working by December 1943. In January 1944, it was successful at breaking a German Lorenz message. A Mark II was immediately ordered with 2,500 vacuum tubes, operational five days before D-Day. Eight more followed before the end of the war, but because of improvements and evolving cryptoanalysis, no two were alike.

Colossus was built as a special-purpose logical computer, but proved flexible enough to be programmed for a variety of tasks, including multiplication.

On March 10, 1943, Charles G Darwin, now Director of the National Physical Laboratory (NPL) [4; 5; 6], informed the Advisory Council of the Department of Scientific and Industrial Research (DSIR) that he was "inclining more and more to the opinion that a Mathematical Department should be established at the National Physical Laboratory." He amplified these thoughts in a paper to the NPL Executive Committee around October 1943. In response, DSIR established an Interdepartmental Committee to "examine the question of whether it is desirable to establish under Government control a Central Mathematical Station" [6]. The Committee had representatives from 11 government agencies, including Hartree and Womersley from the Ministry of Supply, with Darwin as Chairman. On April 3, 1944 the Committee submitted a report to DSIR, stating that the needs of government agencies had been examined and that there was a strong need for a Central Mathematical Station. They recommended that it should constitute a Division at NPL.

The government accepted the concept of continuing to fund computational work at a centralized institution rather than at a number of *ad hoc* locations as during the war. A new Mathematical Division was approved. The official research program for the new Division, in October 1944, included (along with some others):

(1) Investigation of the possible adaptation of automatic telephone equipment to scientific computing

(2) Development of electronic computing devices suitable for rapid computing
(3) Planning and development of components for an improved differential analyzer
(4) Development of a simplified differential analyzer for general use in research laboratories

The first Superintendent of the new Mathematical Division was John Ronald Womersley, probably a nominee of Hartree's who was a member of a Sub-Committee appointed to consider a suitable candidate. The two had collaborated on the application of the differential analyzer to partial differential equations (see Chapter 8).

It happened that NPL was funded by DSIR whose Secretary at the time was Edward Appleton. Ralph Fowler was on the NPL Executive Committee until his death in July 1944, when Hartree replaced him on the Committee. From then on, Hartree played a major "behind the scenes" role in the development of computers.

In February 1945, Womersley went on a two month tour of computing installations in the United States. He became the first non-American to be allowed access to the ENIAC at the Moore School of the University of Pennsylvania and to be informed of a draft report for a new machine, called the EDVAC, whose instructions would be stored in the same way as data. He also saw Howard Aiken's machine at Harvard, the Harvard Mark I, built from the mechanical and electromechanical devices that were being used for punched cards. It was a huge affair with 72 decimal accumulators, capable of multiplying two 23-digit numbers in 6 seconds, and was controlled by a sequence of instructions specified by a perforated paper tape [7; 8].

Before the war, Womersley had learned of Turing machines. These are simple, abstract machines consisting of:

(1) A control unit which can assume any one of a number of finite states
(2) A tape marked off in discrete squares, each of which can contain a single symbol from a finite set
(3) A read-write head that moves along the tape (left or right) and transmits information to and from the control unit.

The *program* of a Turing machine defines the action of the state-symbol combinations. Turing introduced these machines in his paper on *Computable Numbers* [9] in order to study many questions concerning decidability. A *universal Turing machine* is one that can emulate *any* Turing machine.

When Womersley saw Aiken's machine at Harvard, he saw it as "Turing's machine in hardware" [5]. Womersley had initially organized the Division into *four* sections: Mathematical Statistics, Punched Card Machines, General Computing, and Analogue Machines, but after his trip to the US, a new section was created for the design of an automatic computer and Alan Turing was recruited as head of the section.

Just at the end of the war, Hartree too paid an official visit on behalf of the British Government to the USA, where he saw the MARK I at Harvard, and the ENIAC, then still incomplete at the University of Pennsylvania.

On April 22, 1945 in a letter to Slater, he informed him of the impending visit in 3-4 weeks and expressed a desire to see him again:

> I have not drawn up a definite program for the visit, but it seems that most of the people I want to visit and things I want to see on this occasion are in and around Boston and New York; the only exceptions so far are Washington where I have to report to BCSO and get security clearance papers, and Detroit where I want to go to contact some people who have been working on the steering of tanks, which is one of the subjects on which I have done some work recently – it's one of the miscellaneous jobs that has come my way!

He went on to say that he hoped to see his daughter Margaret, who was in the Women's Royal Canadian Naval Service at Halifax, Nova Scotia and now married, but whom he had not seen since 1940.

By May 22, 1945 he was already in Washington. He needed to go to New York and planned to visit Boston for a week. He returned to England on Sunday, July 15, bringing with him five pairs of stockings from Mrs Philip Morse and two packets of bobby pins from Mrs Sam Caldwell for Elaine. Both were wives of faculty at MIT.

As soon as the war secrecy was finally lifted, Hartree published an article announcing the new machine in *Nature* on April 20, 1946. The article conveys his profound excitement at these new developments and describes the machine in simple terms.

> The news has recently been released of a major advance in the development of equipment for extensive numerical calculations, in the successful completion of a large calculating machine based on the use of electronic counting circuits. This machine, known as the ENIAC (Electronic Numerical Integrator and Automatic Calculator), is the invention of Dr. J W Mauchly and Mr. J P Eckert, of the Moore School of Electrical Engineering of the University of Pennsylvania, Philadelphia, where it was

designed and constructed. Its development was sponsored by the Ordnance Department of the US Army and through the interest of Col. P N Gillon and Capt. H H Goldstine. It was designed primarily for the step-by-step integration of the equations of external ballistics, but it includes a flexible control so that it can be applied to many other kinds of calculations within its capacity.

The machine operates by the counting of electrical pulses produced at the rate of 100,000 per second. These are fed to the counting circuits by electronic switching circuits, according to the operation (addition, multiplication, etc.) to be carried out. These circuits count in scale of 10, to ten-figure accuracy. The units in which addition is carried out provide "memory" with a capacity of about twenty numbers, results of previous operations, to which immediate access can be had in the course of further calculations; use of punched cards provides and indefinitely large "memory capacity" for data, intermediate results, or operations instructions, though with much slower access. Final results are delivered on punched cards.

A multiplication table is built into the machine by the connexions of a set of electronic vacuum tubes. The machine also has three function tables, on each of which a set of values of any function required in the calculation (for example, the resistance-velocity relations for a projectile) can be set by hand switches so as to be immediately accessible to the machine which can at any stage read a group of function values and interpolate between them in accordance with an interpolation formula specified by the operation instructions furnished to it. An addition of two numbers held in the machine takes about 0.2 milliseconds; the multiplication of two numbers of 10 decimal digits takes a few milliseconds.

Control of the programme of operations of the machine is also through electrical circuits. The sequence of operations is supplied to the machine by interconnexions made between its different units through plug-and-socket connexions and through hand-set switches, and by coded instructions supplied to a programme control unit; the machine then operates automatically according to the sequence of operations so specified. A certain amount of "judgement" can be included in these operating instructions, in that the machine can be set to select between two alternative courses of procedure at some particular stage of the work, according to the result of a previous operation.

The machine is built up in the form of a number of units, each consisting of one or more vertical panels about 8 ft. high and 2 ft. wide, of which there are altogether *forty*. Each panel carries, at the back, racks of vacuum tubes, relays, etc., and at the front, switches, indicating lamps, plugs, sockets, etc. The different units are interconnected by two sets of lines, one set carrying signals expressing numerical information and the other set for control signals; connexions to and from these sets of lines can be plugged into the various units.

The whole machine comprises about 18,000 electronic valves, 3,000 indicating lamps, and 5,000 switches, and takes about 150 kW in operation.

Its flexibility and speed of operation will make it possible to carry out many numerical calculations, in any field of investigation, which without its assistance would have been regarded as much too long and laborious to undertake.

The ENIAC was the first general purpose automatic computer, built during 1943-45. It was demonstrated to the public in February, 1946.

Hartree also prepared a detailed report for the Executive Committee of NPL. He mentioned the two classes which in US terminology are analogue and digital. In the latter, he identified three subclasses: mechanical counters, electromagnetic relays, and electronic. Many technical aspects were considered. Any member who followed all the details would be well versed in the latest developments.

In the meantime, for Turing it was an exciting prospect to have the British Government support the realization of his Universal Turing Machine. Even before taking up his position at NPL, Turing started thinking about some of the engineering problems. When he arrived at NPL on October 1, 1945 he already was full of ideas for an electronic computer for which Womersley coined the acronym ACE (Automatic Computing Engine).

Turing's first task was to write a report, setting out the detailed design of an electronic universal machine, and an account of its operation. The report, when submitted, contained some surprises. His priorities were a large, fast memory and hardware that would be *as simple as possible*. Logical operations (or, not, and and) were the primitive operations (absent from both the ENIAC and the EDVAC draft design) and, in his philosophy, arithmetic operations could be programmed. Unlike the ENIAC where each decimal digit was represented electronically, the ACE would be binary. Mercury delay lines were recommended for storage. For the logical control, he planned to use some electronic hardware that would contain two pieces of information: where it was on the "tape," and what instruction it had to read there. There were no explicit branching instructions, except that the machine could be programmed to change its own "orders" (instructions). In this way Turing had shifted the emphasis from the building of the machine to the construction of programs.

Although updated later, the ACE report was completed by the end of 1945, an amazing achievement. Womersley wrote a memorandum for Darwin and an introductory report for the Executive Committee meeting February 19, 1946. Turing was asked to explain the principles underlying his proposal. Darwin, as Director, had a number of questions: What would happen if the computer were asked to sum a series that diverged? Could the machine be used for other purposes if it did not fulfill completely Turing's hopes? But Hartree was enthusiastic:

> ... It requires only 2,000 valves against 18,000 in the ENIAC and has a "memory" capacity of 6,000 numbers compared with the 20 numbers of the ENIAC ... if the ACE is not developed in this country the USA will sweep the field ... this country has shown much greater flexibility than the Americans in the use of mathematical hardware.

He ended by saying that the construction of the machine should have priority over his own request for a large differential analyzer.

When the cost of the machine entered the discussion, Womersley introduced the notion of a *pilot* set-up at which point the Committee resolved unanimously to support, with enthusiasm, the proposal that the Mathematics Division should undertake the construction of an automatic computing engine of the type proposed by Turing. But differential analyzers were not forgotten. It was agreed that the ACE had highest priority, then a production model differential analyzer, with the larger differential analyzer at the bottom of the list. The next step with regard to ACE was to present the proposal to the Advisory Council of the DSIR. In this Darwin was successful, obtaining funding for a small, "first-stage" machine with the understanding that, should it fulfill expectations, funds would be awarded for a full-scale version.

Competition also was forming. Newman, the Fielden Professor of Pure Mathematics at Manchester University, the first reader of Turing's *Computable Numbers* paper and co-designer of the Colossus machine, embarked on a plan of acquiring a machine for mathematical problems of an entirely different kind from those so far tackled, such as the four-color problem. In February 1946, he had submitted an application to the Royal Society for a grant to make a start [10]. The Royal Society decided to request the opinion of Darwin and Hartree, both fellows of the Society and knowledgeable about computing. Hartree, in his advisory capacity to the Ministry of Supply, arranged a visit to see the Colossi in order to comment on Newman's application. In the end, Hartree supported the application whereas Darwin did not. The Royal Society then set up a larger committee to investigate the application, consisting of Blackett, Darwin, Hartree, and two mathematicians. Darwin opposed it on the grounds that ACE was to serve the needs of the country. However, he was out-voted on the rationale that the proposed machine was meant for fundamental science, the first machine not required for weaponry. Blackett, who was at the University of Manchester at the time and knew F C Williams from before the war when they had collaborated on the automatic curve tracer for Hartree's differential analyzer, recruited him for the project. Williams was one of the top electronics experts at the Telecommunications Research Establishment (TRE) and had already begun to work on a cathode ray tube (CRT) for display. He was soon joined by others from TRE. Thus the University of Manchester was off to a good start and had a small, prototype machine operational by June 21, 1948 [10] to test the use of the CRT as a memory device, later to be named the "Williams Tube."

In the spring/summer of 1946, at the request of Colonel Paul Gillon of the Ordnance Division of the US War Department, Hartree made a second visit of several months in order to study the ENIAC and its possible application to scientific calculations. He intended also to spend time on behalf of the University, studying Engineering teaching practice, particularly at MIT. Writing to Slater on April 7, 1946, he mentioned his visit, possibly with his wife, Elaine. The Commander Hotel had no vacant rooms but the Caldwells invited the Hartree's to stay with them.

This was an exciting time for the Hartree family. By May 7, 1946, Elaine was already in Ontario, Canada, with her daughter and son-in-law, for the long hoped for reunion, though Hartree was in Washington. Later in the month, June 18-23, he attended the Canadian Mathematical Congress in Montreal. There he presented a paper on *The application of the differential analyzer to the evaluation of solutions of partial differential equations.* The talk was addressed to mathematicians but he stated "I am speaking today from the point of view of the physicist, the chemist, or engineer who is concerned with mathematics primarily as a tool for use in study and application of natural phenomena." He mentioned that partial differential equations were still less amenable to numerical treatment than ordinary differential equations, and talked about his wartime experience with regard to the solution of the heat equation.

He had many friends he wanted to visit. Plans were made to spend time at Cambridge, Massachusetts, to see Slater and the Bainbridges at MIT, van Vleck and Garrett Birkhoff at Harvard, and also the Lindsay's at Providence. Then, to add to the excitement, he was offered the Plummer Chair at Cambridge University. He immediately informed Slater (June 23, 1946):

I have had an invitation to the Plummer Chair at Cambridge (England) which Fowler held. I knew when I came over here that there was a slight chance, about 20% as far as I could judge – of this happening, and before Elaine left (for Canada), I had discussed it with her in case a decision had to be taken while we were apart, and I had decided, if it did occur, to accept. But I was not free to say anything about it, and in view of the probabilities against it, as far as I was aware, I had at that stage to neglect it in making plans for the latter part of any stay here.

A cable with the final offer arrived July 10, 1946. The War Department promised to get reservations for a return journey between July 16-20. He needed to finish his work on the ENIAC (in Philadelphia) and go to Washington to clear final formalities. He still hoped to visit Slater at MIT before his return to England.

A third computer project was soon to be initiated by Maurice V Wilkes at the Mathematical Laboratory at Cambridge, established in 1937 by Lennard-Jones to provide computing facilities for Cambridge scientists. In May, 1946 L J Comrie [11], just back from a trip to the United States, allowed Wilkes to read the *Draft report on the EDVAC* written by J von Neumann. Since there were no copying machines available at the time, he stayed up all night to study it. That summer, Wilkes suddenly received a telegram from the Moore School inviting him to attend a course on electronic computers that was to take place in Philadelphia during July 8 - August 31, 1946 [12]. Before leaving, he had met Hartree who had just returned from the United States and gave him up-to-date information about computer developments there. He also presented him with letters of introduction to Aiken at Harvard and Caldwell at MIT, and suggested that Wilkes get in touch with H H Goldstine who was joining von Neumann's computer project at the Institute for Advance Study at Princeton. This trip had a profound impact on Wilkes, and on the boat on the way home, he planned his own design for an EDVAC like machine, later called EDSAC (Electronic Delayed Storage Automatic Computer). It was a general purpose machine designed with users in mind and was operational by May, 1949.

From Turing's perspective, the design of ACE was completed (having undergone several modifications), and only needed the signal to start. His machine was in the news when, on October 31, 1946, Mountbatten, as president of the Institute of Radio Engineers, gave a speech about the new technology of communication in which he included the remark [5]:

The stage was now set for the most Wellsian development of all. It

> was considered possible to evolve an electronic brain, which would perform functions analogous to those at present undertaken by the semi-automatic portions of the human brain. It would be done by radio valves, activating each other in the way that the brain cells do; one such machine was the electronic numeral [sic] integrator and computer (ENIAC), employing 18,000 valves ...
>
> Machines were now in use which could exercise a degree of memory while some were designed to employ the hitherto prerogatives of choice and judgment. One of them could even be made to play a rather mediocre game of chess! ... Now that the memory machine and electronic brain were upon us, it seemed that we were really facing a new revolution; not an industrial one, but a revolution of the mind, ...

He had taken the information from NPL and the reference to chess-playing machines would suggest that he had heard Alan Turing talk about future possibilities. Darwin and Hartree were embarrassed. They did not want to criticize Mountbatten but each wrote to *The Times* noting that the headline **ELECTRONIC BRAIN** had given a false impression. Hartree's letter to the Editor appeared first, on November 7, 1946. To lend weight to his qualifications, he wrote:

> This summer I had the privilege not only of inspecting but of actually using this machine, and am probably the only person in the country to have done so.

and then expressed his view:

> These machines can only do precisely what they are instructed to do by the operators who set them up. It is true that they can be set up in such a was as to exercise a certain amount of judgment. But it must be clearly understood the the situation in which judgment has to be exercised, ..., must be fully thought out and anticipated in setting up the machine. ...It seems to me the distinction is important and the term "electronic brain" obscures it.

In the past, Hartree generally despised those who wrote letters to the *Times* so his family teased him unmercifully for writing one himself. Darwin's letter followed on November 13, 1946. He took not quite as harsh a point of view, explaining:

> In popular language the word "brain" is associated with the higher realms of intellect, but in fact a very great part of the brain is an unconscious automatic machine producing precise and sometimes very complicated reactions to stimuli. This is the only part of the brain we may aspire to imitate. The new machines will in no way replace thought, but

rather they will increase the need for it, ...

The *Daily Telegraph* was also eager to spread the word, with the headline **BRITAIN TO MAKE A RADIO BRAIN/"Ace" Superior to US Model/ BIGGER MEMORY STORE** on November 7, 1946. Reporters interviewed Womersley, Hartree, and Turing at the NPL. They quoted Hartree as saying:

> The implications of the machine are so vast that we cannot conceive how they will affect our civilization. Here you have something which is making one field of human activity 1,000 times faster. In the field of transportation, the equivalent ACE would be the ability to travel from London to Cambridge ... in five seconds as a regular thing. It is almost unimaginable.

They went on to say that Turing foresaw the time, possibly in 30 years, when it would be as easy to ask the machine as to ask a man. They did admit that Hartree thought the machine would always require a great deal of thought on the part of the operator. He deprecated the notion that ACE would ever be a complete substitute for a human brain. The two men clearly had a different vision. Turing was a programmer whose vision was more in the realm of what humans could do using a computer for a logical process whereas Hartree was the "calculator" (one who computes), seeing mainly what could be achieved with fast arithmetic. His experience had been on a machine designed for arithmetic that was not easy to use.

But little progress was being made in the building of the NPL machine. Hartree asked Wilkes to suggest a collaboration with the ACE project, which Wilkes did, but nothing came of it.

Hartree, as chairman of the Executive Committee's Mathematics Division panel, wrote Darwin in November 1946 stating that the manpower available to ACE was inadequate, comparing it unfavorably with the scale of the US computer projects [6].

When Hartree had visited the Moore School in 1945 he met Harry D Huskey who expressed an interest in working in England. Hartree then arranged for Huskey, one of the ENIAC team, to spend a sabbatical at NPL, joining Turing's group in January 1947. While there he contributed significantly to the ACE project through the design of an ACE Test Assembly, initially approved but not completed [13]. Considerable difficulties needed to be overcome at NPL and consequently opening ceremonies for a "Pilot ACE" were not held until November 1950. By then Turing had left and had a position at the University of Manchester.

When Huskey left, he first went to the National Bureau of Standards in Washington, DC. In December 1948 he transferred to the newly established US Bureau of Standards on the University of California campus in Los Angeles where he designed the SWAC ([National Bureau of] Standards Western Automatic Computer). Construction began on January 1949 and the machine was dedicated in August 1950, some three months before Pilot ACE.

From this brief account of the history of ACE leading to the "Pilot ACE" project, we see that, though Hartree was not ever directly in the line of command, he did all he could, as a member of the NPL Executive Committee and Chair of the Executive Committee of the Mathematics Division, to assure the success of the project, but in an impartial manner in that he would assist others as well. He was never part of the rivalry that existed between the different groups. It was this shared experience between Darwin and Hartree in establishing the Mathematics Division at NPL, which probably explains why it was Darwin who wrote Hartree's *Biographical Memoirs* for the Royal Society about ten years later.

NPL interestingly enough, had not neglected the differential analyzer. In the autumn of 1945, Womersley decided that, as an interim measure, the Mathematics Division should take over the responsibility for the Differential Analyzer Group at the University of Manchester. Hartree was agreeable to the arrangement and, on December 1, 1945, the Differential Analyzer Section of the Division was established at Manchester University with J G L Michel as section head. In November, 1948, the Differential Analyzer was moved from the University of Manchester to a hut called the Babbage Building at NPL. A lot of effort was needed to get this 10 year old piece of machinery back into working order. There it remained until the early 1950's when it was given to the Cranfield Institute of Technology.

Although the Section had worked on the design of a 24-integrator differential analyzer to be built at NPL, it was decided in March 1949 to install a 20-integrator machine being built in Germany. The staff spent much time working on this machine during 1949-54. In an NPL report of May 1954, on *Work of the Mathematics Division*, it was mentioned that a new differential analyzer had been installed, very much like the Rockefeller Differential Analyzer. Results could be presented either in numerical or graphical form, and also input data could be in graphical form. Its mechanical components were connected by electrical servo-mechanisms through a telephone type plug board. The accuracy was limited to 3-4 digits, often sufficient for some practical problems. Some pictures of input tables and the differential analyzer of this professionally engineered machine are found in [6] and on the web. Although this was a vastly improved machine compared to the Manchester one, it was not used much after about two years when it was largely superseded by the digital computer. The machine was dismantled following the decision of the Executive Committee to terminate its use in 1957.

The Manchester machine also soon became obsolete, no longer meeting the high technologoy needs of Cranfield Institute. In 1965 it was sent back to Manchester University where it was photographed, dismantled, and carefully stored in packing cases. For nine years they remained in obscurity. Then in 1974, the Science Museum of London discovered the machine's existence and commissioned its reconstruction. A four-integrator machine is presently on display at the Science Museum in London. The remaining four-integrator machine is located at the Manchester Museum of Science and Industry.

Spreading the news

After his articles in Nature had appeared, not to mention the publicity associated with his letter to the London Times, Hartree's advice was sought by many groups. In a letter to Bohr on March 9, 1947 he mentioned that he had received an invitation from the Swedish National Council for Scientific Research to go to Stockholm and give some lectures on recent developments in calculating machines and expected to go March 20. He also was invited to give lectures in Lund about a week later and would like to see Bohr in Copenhagen, if he was free.

As an OB parent, Hartree presented a lecture on Calculating Machines at Bedales on Parent's Day. An account in the Bedales Chronicle [15], mentions how fortunate they were to have the foremost authority on the subject in the country and one who had recently visited America, talk to the school. Hartree had begun by showing them a simple desk calculator that looked like a typewriter. He then explained the huge and complicated, but very fast, ENIAC built in the US. He presented slides of photographs to dispell any misconceptions of what this "machine" might look like. The students found his talk interesting in that it succeeded in telling them a great deal without going beyond the comprehension of the non-mathematical.

Industry too was getting interested in electronic computers. Britain's leading caterer of tea, coffee, restaurants, and tea shops was J Lyons and

Co., with headquarters in London. The business consisted of an enormous volume of small transactions requiring an immense amount of paperwork. The company was very advanced in administrative methods and had recruited Cambridge mathematicians already in the 1930's to assist their organization. Having heard about the ENIAC, management approved a study tour to the US and Canada in May/June of 1947. Ironically, the representatives never got to see the ENIAC but they learned about a more advanced, but similar project at Cambridge University being developed by Maurice Wilkes [16; 17]. Their first stop had been a visit with Herman H Goldstine, then at the Institute for Advanced Study, Princeton University. He had not previously considered the possibility of an electronic computer in a commercial office but immediately became interested and began to outline how such a project could be tackled. He asked them to return later so he could help them futher which they did. He also wrote immediately to Hartree. Within a week, Lyons & Co. received a letter from Hartree asking that representatives come to see him and Wilkes at the Cavendish Laboratory in Cambridge. Immediately upon their return, they arranged for such a visit. A report of both visits was submitted to the Lyons Board on October 20, 1947 [17]. It must have made a favorable impression, for J R M Simmons, Chief Comptroller, wrote a note in his diary on October 23, 1947 [18] that Mr Harry Salmon, Chairman of the Lyons Board of Directors, was prepared to recommend a donation to the Mathematical Laboratory, without strings attached, since he thought "it would help Hartree to further his research." He also was contemplating spending £50,000 to £60,000 on the building of a machine, recognizing that a second would also be needed because of the risk of breakdowns. A more formal visit with a delegation of six representatives occurred on November 11, 1947. A very favorable report of this visit includes the remark [18]:

> Neither Professor Hartree nor Dr. Wilkes appear to be scientists who are only interested in scientific speculation and are careless of the time taken to achieve some result. Nor are they dogmatic in their opinions and they seem willing to co-operate with others, even laymen like ourselves.

Hartree and Wilkes admitted they were primarily interested in scientific possibilities, including a machine Hartree needed in his own work. At the same time, they were interested in commercial applications because it would demonstrate the machine's versatility. Hartree had also pointed out that his complicated calculations and their simpler ones were fundamentally the same in that both were a combination of simple addition and multiplication. At the same time, they were willing to share their knowledge and believed they were closer to producing a complete machine than any others in the field. When asked whether the loan of an electrician from the Company would be of assistance, they agreed readily to the proposition since it was not easy to get a competent person with uncertain permanent employment. Thus a strong relationship was started. A donation of £3,000 was soon made to the Mathematical Laboratory along with the services of an assistant. This support came just at the right time when momentum for the EDSAC was slowing down due to a lack of funds.

Even before EDSAC was completed, discussions were underway for the design of the world's first electronic business computer, LEO (Lyons Electronic Office). In 1951 it began to carry out time critical, error-free business accounting. Though closely based on EDSAC there were significant differences. The system started running the Lyons payroll in 1954. J Lyons & Co. went on to create a separate company, LEO Computers, that supplied machines and a consulting service. It was housed in new offices in the upper floors of the Whiteley Department Store in Queensway, London W2. After Hartree's untimely death in 1958, the company asked if the London offices could be named after him. The family readily concurred and the Offices became known as "Hartree House" in recognition of the support and encouragement Hartree had given the Lyons Board in the early days.

Notes and References

[1] Porter A 1938 (private communication)
[2] Blackett P M S and Williams F C 1939 *Proc. Camb. Phil. Soc.* **33** 494
[3] An excellent overview of computing in Britain including all forms of machines is presented by Mary Croarken 1990 *Early Scientific Computing in Britain* (Oxford: Oxford University Press)
[4] The National Physical Laboratory was founded in 1900 as a government agency to bring scientific knowledge to bear in industrial and commercial life and break the barrier between science and commerce. At the opening ceremony on March 1902, the Prince of Wales said "...the nation is beginning to realise that, if its commercial supremacy is to be maintained, greater facilities must be given for furthering the application of science to commerce and industry." Sir Charles Galton Darwin was director of NPL, 1938-49, succeeding Sir William Lawrence Bragg 1937-38. Today it is the United Kingdom's national standards laboratory.
[5] Much of the information about the ACE project is given in the book, Hodges A 1983 *Alan Turing: The enigma* (New York; Simon and Schuster). Scanned images of many of the minutes of the meetings of the NPL Exec-

utive Committee, Turing's report, and other information, can be viewed at the site
http://www.alanturing.net/archive/index/aceindex.html

[6] Yates D M 1997 *Turing's Legacy: A history of computing at the National Physical Laboratory 1945-1995* (London: Science Museum)

[7] An excellent general source of information about computers is *Encyclopedia of Computer Science*, edited by A Ralston, E D Reilly, and D Hemmendinger, (Oxford: Nature Publishing Group) Fourth Edition 2000.

[8] Burks A W and Burks A R (1981) *IEEE Annals of the History of Computing* **3** No 4, 310; also Barkley Fritz W 1994 *IEEE Annals of the History of Computing* **16** No 1, 25

[9] Turing A M 1937 *Proc. Lond. Math. Soc.* **42**

[10] Croarken M 1993 *IEEE Annals of the history of computing* **15** No 3, 9 outlines scientific computing in Britain during and after the World War II. Then the period 1946-48 is discussed as the start of computer research at Manchester University.

[11] Leslie J Comrie (1893-1950)) was one of the few kindred spirits with Hartree. His main goal in life was to "spread the gospel of mechanical computation." He became a Fellow of the Royal Society just before his death. "L J Comrie", *Obituary Notices of Fellows of the Royal Society* **8**, 100 was written by H S W Massey and appeared in 1952.

[12] Wilkes M V 1985 *Memoirs of a Computer Pioneer* (Cambridge, MA: MIT Press)

[13] Huskey H D 1984 *IEEE Annals of the History of Computing* **6** No 4, 350

[14] Marsh P 1978 *New Scientist* **80** No 1134

[15] Gurney I 1947 *Bedales Chronicle* p 15

[16] Bird P J 1994 *LEO: The First Business Computer* (Hasler Publishing)

[17] Caminer D, Aris J, Hermon P, and Land F 1998 *LEO: The Incredible Story of the World's First Business Computer* (New York: McGraw Hill)

[18] Simmons J R M 1947 *Electronic Machines* Modern Records Centre, Library, University of Warwick

Chapter 14

Returning to Cambridge

In 1929, the John Humphrey Plummer Foundation was established at Cambridge University to manage the funds left under the will of the late Mr Plummer "on perpetuity for the promotion of education in Chemistry, Biochemistry, Physical Science or other allied subjects in the University of Cambrige ... and endow ... Professorships each of an annual value of £1,200." Mr Plummer, who died on December 26, 1928, had lived in Southport, Lancashire for fifty years where he had established a very successful real estate business. Upon his death, many of his assets were tied to leased property with long leases remaining. Thus his estate had the ability to earn income over the years estimated at £10,000 per annum. There is no record that Mr Plummer had any association with Cambridge University nor any special interest in science. His bequest seems to have arisen from a genuine desire to use his wealth in the best possible manner for the benefit of mankind, though the timing of the will, written in 1921, suggests that the death of his youngest son in World War I may have been a factor. His wife predeceased him and only his elder son – a church organist in Derby – survived him [1].

In 1931, the first three Professorships were awarded to Ralph H Fowler in the area of Mathematical Physics, to E K Rideal in Colloidal Science, and J E Lennard-Jones in Theoretical Chemistry. With the death of Ralph Fowler on July 28, 1944, Cambridge University had no theoretical physicist associated with the Cavendish Laboratory though Dirac was Lucasian Professor of Mathematics. According to Mrs Hartree, Cambridge students with problems would write to Hartree, who in 1937 had been promoted to Chair of Theoretical Physics at Manchester University, a position Blackett, who succeeded W Lawrence Bragg as the new Langworthy Professor of Physics, had persuaded the University to create. Other sweeping changes occurred. With remarkable speed, Blackett created a major research center for cosmic rays. After the war, Hartree's interests had shifted more towards computing. According to the University of Manchester Registrar, at the May 11, 1945 meeting of the University Council "it was resolved that Professor D R Hartree at present Professor of Theoretical Physics, be invited to accept an appointment as Professor of Engineering Physics in the Department of Engineering, as from September 29th, 1945." This may have been arranged by Blackett so he could create a Chair for Léon Rosenfeld, a distinguished nuclear physicist, as Professor of Theoretical Physics. In this he was successful – Rosenfeld took up his duties on June 20th, 1947. One can speculate that it also suited Hartree. We know that Blackett supported M H A Newman in applying for a grant from the Royal Society to cover the cost of building and staffing a computing machine.

After the war, Cambridge University decided to continue the Plummer Professorship in Mathematical Physics within the Faculty of Mathematics, and began looking for a successor to Fowler. They offered the position to Nevill Francis Mott [2], who had started research at Cambridge under Fowler. At the time, he was doing well at Bristol with a large solid state group, well supported financially and morally by the university. At Cambridge he sensed he would not have this support and would have to share a secretary with Lawrence Bragg who was now the Cavendish Professor. In succeeding Rutherford, Bragg was having a difficult time establishing his own priorities in research, a strong anti-Bragg faction wanting to continue nuclear research. For these reasons, Mott declined the offer. The Plummer Professorship was then offered to Peierls, who also declined, doing very nicely at Birmingham where he was just starting work on nuclear reaction theory. Offers to Casimir and Kramers fared a similar fate. It was then offered to Hartree who was visiting the Moore School of Engineering, Univeristy of Pennsylvania, Philadelphia, at the time.

According to Mrs Hartree, the family had very much enjoyed the non-academic city of Manchester, but the ties to Cambridge were strong and possibly Hartree saw the Cambridge position as one with fewer responsibilities permitting more time for research and travel.

One should remember that as early as February 1946, Wilkes was settled as director of the Mathematical Laboratory at Cambridge. In his view, the laboratory's role was to provide equipment with sufficient advice and technical assistance for Cambridge scientists to do their research and in a laboratory report had written [3]:

There is a big field here, especially in the application of electronic methods, which have made great progress during the war, and I think Cambridge should take its part in trying to catch up some of the lead the Americans have in this subject.

With the support of the University, he was in a position to initiate computing machine research. Hartree was well connected with Cambridge and, according to Wilkes [4], must have been aware of what was happening.

According to Arthur Porter, whom Hartree consulted for advice, it was a difficult decision, but Hartree accepted the position as Plummer Professor of Mathematical Physics and returned to Cambridge. By the time he arrived, Wilkes was determined to build an electronic stored program computer. Furthermore, the laboratory had sufficient funds to get a project started without requiring permission or funding.

Hartree's appointment was announced in the "News and Views" section of *Nature* on August 17, 1946, mentioning that his "appointment to the Cavendish Laboratory will bring him in touch with many fields of physics where his mathematical and computational ability will find full scope."

The picture in Figure 14.1 shows Douglas Hartree in his new office at the Cavendish, wearing what was then considered a large, garish tie, one of several that he had brought back from America. His family did not approve but the students thought they were fun.

Though not required to do so, Hartree gave an inaugural address in October, 1946 to the Cambridge community, entitled *Calculating Machines: Recent and Prospective Developments and their impact on Mathematical Physics*. He said he selected the subject because it was one in which there already was a great deal of interest in the United States. He believed that much more would be heard of it and that it was time that physicists and others began looking ahead and thinking about what they would like to do when the equipment became available. He then proceeded to give a general description of the ENIAC, which he had used just a few months earlier.

In the description of the "master programmer", a unit that controls the repetitive execution of a computing sequence, he made quite clear that use of the machine did not replace the thought needed for performing a calculation, only the labor. To him, this was a point of great importance and one that he believed was missed entirely by those who spoke of an "electronic brain." One of the points he made, was the importance of taking a "machine's-eye view" of a problem. He used the example of dividing by zero. A human would *not* divide by zero whereas a computer would take

Fig. 14.1 Douglas Hartree in his office at the Cavendish wearing a tie brought back from America.

what ever action it was designed for.

Hartree was impressed with the speed of about a million multiplications per hour. He thought the limited high-speed memory capacity of 20 accumulators was the most severe limitation and in this he was correct. He then went on and took the dangerous step of mentioning independent estimates of needed memory capacity as being between 1000 and 5000 numbers! In retrospect, this statement probably should have been qualified by a remark like "for machines of this speed ..."

In discussing the impact of these developments on mathematical physics, he correctly identified things that needed to change. New methods were required and an understanding of their limitations. As an example of the former he mentioned systems of equations, where standard text-books in algebra showed the solution in terms of determinants:

> Have you ever tried to evaluate a determinant of order 50? If you have, you soon will wish you had not!

Using a system of three equations in three unknowns, he showed how this problem could be "reduced" to a minimization problem which might be easier to solve on a computer with limited storage than the system of equations. In another example, he pointed to a new field of research on the theory of iterative methods.

The one problem he did *not* mention was the effect of round-off when so many operations were performed. It may be that he was so accustomed to working with 5 or 6 digits, that having a machine that worked with 10 digits was enough to protect against the accumulation of round-off.

It was only a few weeks after this address that *The Times* article, mentioned earlier, appeared in connection with the ACE project referring to an "**Electronic Brain**." It is clear he could not avoid being caught in the controversy (see Chapter 13).

As soon as time permitted, Hartree gave high priority to writing and publishing papers that had been set aside because of the war.

At the outbreak of war his father's calculations had been sent to Slater, who delegated the task of preparing them for publication to Dr Millard F Manning. But Manning had died in 1942 leaving some results still unpublished. Hartree's own father died in 1943. In early 1947, in memory of his father, Hartree wrote up the work on *Self-consistent field, with exchange, for nitrogen and sodium*. All the terms of neutral nitrogen (^4S, ^2D, ^2P) were considered, but omitting superposition of configurations. The next paper was on sodium (Na), the first atom for which solutions of Fock's equations had been evaluated through pioneering work by Fock and Petrashen [5]. They evaluated the equations using a method based on the use of Green's functions. While formally correct, this was hardly the most suitable method for practical computation and, in comparing wave functions with earlier work on Na^-, Hartree and Hartree had found considerable irregularities and a greater difference in energy than expected. They performed their own calculations and Hartree published the new wave functions. The paper was submitted on June 20, 1947.

Shortly thereafter, on September 9, 1947, Hartree's mother, Eva Hartree, died at the age of seventy-three. Of the five children born to her, only Douglas Hartree survived. It is not difficult to imagine the special bond that must have existed between them. She was a dynamic person, full of kindness, known to have helped many through her social causes, yet was embarrassed by gratitude for her work. At the same time, she was a strong-willed person. Already in the 1920's she was driving a car. She made certain that Elaine too was properly trained to change a tire and knew how to hand crank an engine in order to get it to started. Both skills were needed quite often in those days. Her interest in women's issues led her to emphasize self-reliance. She believed wives should not be totally dependent on their husbands in all matters. This together with the Hartree intellectual approach to problems, trying all possibilities before seeking help, led to a culture of "self-reliance", somewhat different from Samuel Smiles "self-help," at least in the opinion of her grandson, Richard Hartree.

On returning to Cambridge, Douglas and Elaine had moved into a house on Barrow Road, the next one after Bentley Road, away from town. His mother was then living at 21 Bentley Road in the house that had been built for them when they first were married. Upon her death, they moved back into the house, but now without live-in domestics. When Elaine was asked whether she would have preferred live-in domestics to doing the work herself (with occasional help from a cleaning lady), she claimed definitely to prefer doing so herself and not having to be concerned about the feelings and foibles of live-ins, not to mention organizing them. To help make life a bit easier, and possibly because of a love for gadgets, Douglas had purchased a Thor washing machine for her in America that he had shipped to England. It came complete with interchangeable tubs, so that it could be used for both clothes and dishes. Douglas was great at drying dishes that had been washed by hand and was the "official stacker" of the dish washing machine.

Another major paper was the one on *The laminar boundary layer problem in compressible flow.* This was the problem he had attempted to solve of the ENIAC as a "benchmark calculation" and where he had failed to take "the machine's-eye view" that made a profound impression on him. Together with W F Cope, a member of the Engineering Division of NPL, a detailed account of this work appeared in 1948. The work was carried out on behalf of the Chief Scientific Office, Ministry of Supply.

Historically, the work arose from an attempt to estimate, if only qualitatively, the effect of a boundary layer on the aerodynamic force coefficients of a projectile, a study that emphasized the dearth of information, and from an attempt to calculate the position of separation of the boundary layer as a function of the Reynold's number and Mach number. Some photographs had been obtained from wind tunnels but the problem had not been investigated theoretically. The paper consisted of three parts. The first was a general survey and a derivation of equations (written by Cope), the second described an algebraic preparation of the equations to be solved, and the third described the solutions of these equations on the ENIAC and presented the numerical results.

In many respects, this paper has one of the best descriptions of the use of the ENIAC from a programmers point of view, including diagrammatic representation of the master-programmer connections for an iterative method of computing the quantity $(1+\alpha r)^{1/9}$ and the two "steppers" that were needed. To someone like Hartree, familiar with setting up a differential analyzer for a specific problem, the task of setting up the necessary connections for the machine would not be as daunting as they may seem to the reader.

Most of the numerical work had been carried out by Hartree in the summer of 1946 during the course of his visit to the Moore School of Electrical Engineering of the University of Pennsylvania, at the invitation of the Ordnance Department of the US War Department. He expressed his deep appreciation for the opportunity of acquiring first-hand experience on the ENIAC. He thanked Colonel P N Gillon, Office of the Chief of Ordnance, for making the arrangements for the visit; Dr L S Dederick, Ballistics Research Laboratory, Aberdeen, Maryland for making the ENIAC available for the work described in the paper; Dr and Mrs H H Goldstine, Dr D H Lehmer and other members of the group engaged in the operation of the machine (especially Miss Kay McNulty) for instruction, advice, and help in organizing the machine, planning the machine set-up for it, and running the machine. He concluded:

> The active and friendly help received made the work, in addition to being of absorbing interest, a real pleasure.

The solutions obtained from various iterative methods were evaluated after his return to Cambridge.

His last major paper in 1948 got him back into thinking about atoms. Over the years, Hartree in collaboration with others (not necessarily coauthors) had published a number of articles but there was none that represented the current state of atomic structure calculations. In 1948, he published a major paper, *The calculation of atomic structures*, in *Reports on Progress in Physics*. It was a comprehensive review both of theory and of practice. It contained the memorable paragraph quoted in his *Biographical Memoirs*, and used by many others to this day (the italics are added for emphasis).

> The problem of calculating the atomic structure of an atom or ion, other than one with just a single electron, is the problem of solving the wave equation for a many-particle system. This equation has no exact formal solution in finite terms, and it is necessary to use approximations both for this reason and for a more practical one which is this: for an atom of

p electrons, the solution is a function of $3p$ variables, and even if it were possible to evaluate such a solution to any degree of numerical accuracy required, no satisfactory way of representing the results, either in tabular or graphical form, is known. *It has been said that the tabulation of a function of one variable requires a page, of two variables a volume, and of three variables a library;* but the full specification of a single wave function of neutral Fe is a function of seventy-eight variables. It would be rather crude to restrict to ten the number of values of each variable at which to tabulate this function, but even so, full tabulation of it would require 10^{78} entries, and even if this number could be reduced somewhat from considerations of symmetry, *there would still not be enough atoms in the whole solar system to provide the material for printing such a table.*

Hence the need for finding suitable approximations.

The paper then reviewed the theory, starting with wave functions in the form of determinants of one-electron wave functions (interestingly enough, he did not refer to them as Slater determinants – a term that must have been introduced later) or linear combination of determinants that satisfied Pauli's principle. He discussed the derivation of the energy expression needed for deriving "equations with exchange," as he called them. Computational methods were outlined briefly, followed by a section entitled *Improvements in the approximation.* This section exhibited his deep understanding of the field.

The last section was simply *Results.* This section included the only table of the paper. It was a table of references to *all published results* as a function of the atomic number and degree of ionization. Entries indicated the authors and the method used. Of course, such a table is unthinkable today since a Hartree-Fock calculation for even a heavy atom can be performed in a fraction of a second on most computers and with much higher precision. He then went on to mention special cases not included in the table, namely negative ions, high states of ionization, states of high excitation, and finally states of the continuum. In the latter, the energy parameter was given but the fact that the wave function remained finite required a different treatment at large r. A possible treatment had been published by Hartree, Kronig, and Petersen in 1934.

The review article was extremely valuable to those wishing to perform new calculations or others needing wave functions. But if these were not available, Hartree could be called upon for help. Professor Michael J Seaton, University College, London, recalls that in about 1949 he had a need for some atomic wave functions which were not included in the *Reports on*

Progress in Physics article. David Bates, his supervisor, said that Hartree might have them and suggested that he write to him. He did and Hartree replied that he would calculate them. Some 10 days later he sent what had been requested. Seaton was most impressed that the great man should do all that work just to oblige a first-year research student!

When Hartree had an occasion to write to Bohr, he mentioned that since 1939 he had been engaged in purely classical work and was now relearning quantum mechanics, though it was coming back more easily than he had feared it might. With this review he was back in stride.

Upon his return to Cambridge, Hartree rejoined the Cambridge Philharmonic (Town) Orchestra. He had an excellent sense of the structure of music, a good ear, and enjoyed playing tympani. Orchestras usually did not have their own sets of tympani. Hartree purchased a set for the Cambridge orchestra so he could play at rehearsals as well as during performances. Roy Garstang [6], one of his students, informed me:

> I have a recollection of a concert in a church in Cambridge. It was at the Wesleyan Methodist church and it was "Messiah." The date was December 3, 1949. This was just before the end of term and I was a first year research student. Hartree was playing the tympani, obviously enjoying himself greatly, and looking as red as a strawberry with the effort he put into it!

Notes and References

[1] Hartree J R 2000 (unpublished)
[2] Mott, Sir Nevill Francis 1987 *A Life in Science* (London: Taylor & Francis)
[3] Croarken M 1992 *IEEE Annals of the History of Computing* **14** No 4, 10 discusses the emergence of Computer Science research and teaching from the mid 1930's to about 1949.
[4] Wilkes M V 1985 *Memoirs of a Computer Pioneer* (Cambridge: MIT Press)
[5] Fock V and Petrashen M 1934 *Phys. Z. Sowjet* **6** 368
[6] Garstang R H 1998 (letter of July 8)

Chapter 15

Summers in North America

In the United States, it had been recognized that large-scale automatic digital machines would make more demands on individuals than vice versa. In other words, skill in the analysis, formulation, and programming of problems would become the controlling factor in the proper use of these machines. To attack this problem, the Institute of Numerical Analysis of the National Bureau of Standards was established early in 1948 on the campus of the University of California in Los Angeles. Hartree served as Acting Director for three months during the summer of that year.

Shortly after the Institute was established, a series of symposia were held on the development of high-speed automatic computing machinery and related numerical methods. Most of the papers presented were in the nature of progress reports on various projects in the United States and Great Britain, but the last symposium, held on July 29-31, dealt with the future of numerical analysis.

Four of the papers presented at the mathematical sessions were collected later into an official government document, *Problems for the Numerical Analysis of the Future*, National Bureau of Standards, Applied Mathematics Series 15 (1951). The first paper, *Some Unsolved Problems in Numerical Analysis*, was by Hartree. He stated:

> It is, of course, rash to talk of "unsolved" problems in numerical analysis; what I really mean, of course, is problems to which I do not have the answer. What I propose to do is to present a series of questions, not concerning large, spectacular problems like the prediction of weather by numerical integration of the equations of motion in the atmosphere – which is a possible problem in numerical analysis – but a number of much smaller questions, ones that I have come across in the course of my own work; problems to which the answers should be known, ... before the larger problems are tackled.

He then mentioned the problem of eliminating approximately known roots from a polynomial. His second question concerned the solution of systems of simultaneous non-linear algebraic equations. In this section he touched upon a topic that in Computer Science today is called "complexity of algorithms." His observations indicate great insight on his part:

> Whether a method practical on a small scale will also be practical on a large scale depends very much on the way in which the number of operations needed increases with the number n of equations; whether it increases as a power of n ... or exponentially with n (as 4^n, 5^n, ...) or perhaps $2^{2^{2^{2^n}}}$ . .... For large n it matters very much if the number of operations increases as n^4 or 4^n.

The final problem he mentioned was the psychological one. Here he talked from personal experience:

> One of the unsolved problems of numerical analysis is how to overcome the attitude of the mathematical fraternity towards the subject – an attitude exemplified by the comment of a distinguished mathematician, introducing a lecture of mine on the mechanical integration of differential equations, that he had always regarded the solution of differential equations as "a very sordid subject."

This is the only time in his writings that Hartree gave an indication that his computing achievements did not always result in recognition and admiration. In 1963, after Hartree's death, Nevill Francis Mott (1905-1996) was interviewed by T S Kuhn for the Archives for the History of Quantum Physics, with a copy stored at the Center for History of Physics, American Institute of Physics. When Mott was asked about Hartree during his student days, he really did not remember very much, but went on to say:

> ... his contribution was just that single thing, really – the self-consistent field. Calculating one atom after another. He used to say "Charming, you know; I really like arithmetic." And all the rest of us absolutely hated it! I don't think you could say he did make much contribution except as a teacher and a teacher at the undergraduate level, really.

Insight may be gained from Slater's remarks [1]:

> Douglas Hartree was very distinctly of the matter-of-fact habit of thought that I found most congenial. The hand-waving magical type of scientist regarded him as a "mere computer."

After Hartree had completed his appointment as Acting Director at UCLA, he visited the University of Illinois in Urbana, at the invitation of Dean Louis N Ridenour, Dean of the Graduate School, to give a short series of lectures on calculating instruments and machines. The lectures were devoted primarily to recent developments but Dean Ridenour had asked him to give some attention to the historical side.

The first part of the lecture series was devoted to instruments, chiefly the differential analyzer with its integrator device. It included a discussion of the new Rockefeller differential analyzer that had been built by Bush and Caldwell [2] where the assembly of cross and longitudinal shafts and cross-drives for interconnections between them were replaced by electrical systems, making them much easier to use.

As examples of the application of the differential analyzer, Hartree selected some of his most challenging problems that had arisen during the war, namely partial differential equations. In the parabolic heat conduction problem, he showed how "Method I" was obtained by replacing the time derivative by a centered difference and compared it with "Method II" obtained by replacing the second-order space derivative by a finite difference (see Chapter 12). An important difference was the role of boundary conditions in the two cases. Then he went on to the hyperbolic equations of which the wave equation in a uniform medium with a constant velocity of sound was an example. A straight-forward approach was beyond the capability of the differential analyzer but he showed how the problem could be solved by the method of characteristics on the differential analyzer. He then discussed the problem of supersonic flow past a shell, an example of a laminar boundary layer problem (Chapter 10).

The lectures dealing with an introduction to large automatic digital machines were of particular interest at this time. He started this chapter with some preliminary notes on terminology that are worthy of note in that they showed the care he devoted to language:

> Any substantial scientific development involves building up a terminology, either of new words or of current words used in new or specialized senses, appropriate to the special features of the new development. There are disadvantages both in coining new words and in using current ones in special senses. Too free a use of new words gives the impression that the new development is something esoteric, closed except to those who have been initiated and learned the secret passwords, whereas specialized use of words already current may lead to misunderstanding, particularly when words habitually used in connection with living organisms, and especially with human activities such as "memory," "choice,"

"judgement" are applied to mechanisms.

...I have preferred to make use of current words rather than to invent new ones, and to use them in a way not specially related to that of any one group. A few terms need some comments. "Instruction" is used, in preference to "order," for the statement of an operation the machine is required to carry out. This use of the longer word is preferred because "order" has another meaning which may be required in the same context, as one speaks of *taking the operating instructions in order*. "Programming" and "coding" are apt to be used as almost synonymous. I have preferred to make a distinction between them, using "programming" for the process of drawing up the sequence of operations for a particular calculation, and "coding" for the process of translating either numbers or instructions into the form in which they are supplied to the machine, or in which they occur in the machine. In this sense, a number or instruction can be "coded" on an IBM Card or to a set of relays. A calculation must be "programmed" in terms of a set of instructions which can be "coded" in accordance with the facilities of the machine. "Control" seems to be used almost indiscriminantly for the operation of taking instructions in the appropriate order and initiating the appropriate action on each one, and for the part of the machine which carries out this operation. I have tried here to use "control" for the operation, and "control unit" or "control system" for the hardware which carries it out. I have preferred to use the neutral word "storage" or "store" rather than "memory" but have found no satisfactory neutral word to take the place of "judgement:" the alternatives "choice," "selection," "discrimination" that have have been suggested seem unsatisfactory, and a more comprehensive term is required to express the faculty involved here. For the selection of an instruction out of the normal sequence, on the basis of the assessment of one or more criteria, I have introduced the term "conditional selection" in place of the "conditional transfer" which is sometimes used, as without reference to a particular machine it is not obvious that any "transfer" is involved.

Much of Hartree's suggested terminology has been adopted though, particularly at Cambridge, it has long been the custom to refer to "machine orders." The processes of making "selections" are now often referred to as "decision processes" and "judgement" is rarely used. Programming still is considered a more high-level task than coding.

Several chapters were devoted to an elementary introduction to computers, including a description of the ENIAC. Here Hartree elaborated on how easy it is for a calculation to go wrong. The laminar flow problem required the evaluation of $1/(1 + \alpha r_0)^{1-\beta}$ for which an iterative process was used. An initial approximation was to be obtained from a function table so that only two iterations would be needed. Here it was assumed that r_0 would always be positive. But r_0 was also computed, and what had not been anticipated was that, in one of the trial solutions r_0 might become negative. This in itself need not have caused a severe problem except that on the ENIAC negative numbers were represented in a 10's complement form so -0.01 was interpreted as 0.99 and accurate values could not be obtained in only two iterations. This was a situation where a human would have exercised some intelligence whereas the machine, not being programmed for negative values, did something sensible according to its structure and spurious results were obtained.

The moral of this experience for Hartree was that, in programming, it was necessary to look at a series of instructions from the point of view of the machine which would follow them literally. He ended by saying that this was not as easy as it sounded since it was quite difficult to put oneself in a position of doing without *any* of the hints that intelligence and experience would suggest to a human computer in such situations. He did not foresee the possibility of using the computer itself as an extension of the human to assist in what is now called debugging.

The rest of the lectures dealt with future developments. He mentioned new high-speed storage systems such as acoustic delay lines, magnetic material either on a tape or drum, insulating screens with electrostatic storage. All were under development at the time. The ENIAC had been a decimal machine, but Hartree mentioned the possibility of "binary" machines, though he did not use that term. He described Boolean algebra and the logical design of an adder for binary numbers, and ended with a discussion of programming and coding. Here he introduced the "flow-diagram" that von Neumann and Goldstine had proposed [3]. It should be remembered that Hartree was discussing machine language programming without the aid of compilers or even interpreters.

His final lecture was on numerical analysis for high-speed automatic digital machines. Here he defined the process of iteration, so fundamental to the use of computers. He mentioned that practical methods for simultaneous algebraic equations could be classified as "direct" or "indirect," where the latter were essentially iterative methods. He reported that direct methods had come under suspicion on account of some estimates of possible rounding errors, but that work in the Mathematics Division at NPL had shown that for many applications these estimates were far too pessimistic. He then posed the algebraic problem as one of minimizing the sum of squares of residuals. Today there is an extensive literature on the method of steepest descent or gradient methods. Hartree commented that from his experience with small systems on the differential analyzer, the method was not as straightforward as it sounded. The final part of the lecture made a brief mention of ordinary and partial differential equations, but perhaps time permitted him to only touch on these problems briefly.

The set of lectures given at the University of Illinois were published by the University of Illinois Press at Urbana in 1949. Hartree repeated these lectures at Cambridge, adding references to the EDSAC machine then being built at the Mathematical Laboratory. At his request, Wilkes invited him and his audience to go over to the Laboratory to see the machine for themselves. A revised book of lectures was published by the Cambridge University Press in 1950.

The visit to America in the summer of 1949 was a shorter one in which he collaborated with Garrett Birchoff, an applied mathematician at Harvard who had been Fowler's student. During the postwar years, 1945-50, Birkhoff devoted most of his time to fluid mechanics. In 1949 he used Aiken's Mark I calculator at Harvard for the study of jets, waves, and cavities [4]. The computing was organized by Hartree and others. In this research, Birkhoff was supported by the US Office of Naval Research with a grant that made possible repeated visits to Harvard by Hartree and others.

On this trip, Hartree was accompanied by his wife and his son, Richard, who recalls time spent at the Hazen's cottage on Lake Winnipisaukee, New Hampshire, and a trip to Philadelphia where they visited the ENIAC project. As tourists they visited Washington, New York, and the Lindsays at Brown University. Writing about their trip to Slater, Hartree said (December 2, 1949):

> We had enough of the heat to realize fully why people like to get out of Boston and Cambridge for the summer if they can!

Thus it appears the Slaters were on vacation.

Notes and References

[1] Slater J C 1975 *Solid-State and Molecular Theory: A Scientific Biography* (New York: John Wiley & Sons)
[2] Bush V and Caldwell S H 1945 *J. Franklin Inst.* **240**, 255
[3] Goldstine H H and von Neumann J 1947 *Planning and coding of problems for an electronic computing instrument* (Institute for Advanced Study, Princeton; duplicated)
[4] Birchoff G 1990 *A History of Scientific Computing* Edited by S G Nash (New York: ACM Press) pp 63

Chapter 16

Mathematical Laboratory, numerical analysis, and teaching

As Plummer Professor, Hartree had an office in the Cavendish Laboratory, and though he had no official duties in the Mathematical Laboratory, he showed a keen interest in its activities and generously offered his services whenever he thought he could contribute.

The EDSAC became operational on May 6, 1949. Soon thereafter Maurice V Wilkes, David J Wheeler, and Stanley Gill, wrote a book entitled *The Preparation of Programs for an Electronic Digital Computer* that was published in 1950 [1]. This was the first programming book written with users in mind and included a number of important ideas such as the use of library subroutines. A supplement, *Introduction to Programming for an Automatic Digital Calculating Machine and User's Guide to the EDSAC*, appeared in 1954. The first part was written by Hartree and was an introduction to programming that gave a fuller treatment, with more examples, of the earlier chapters covered by the book.

The first users of the EDSAC were members of the Laboratory, staff, students and, with one exception, Douglas R Hartree. According to Wilkes [2]:

> His knowledge of numerical analysis and, more particularly, his wide experience of computing of the most practical kind, qualified him to play a major role at that critical moment. He had personal qualities of no less value. He was entirely without any sense of his own importance and could work, seemingly on equal terms, with those much younger than himself. He never attempted to lay down the law and detested those who did. ...It was, in no small measure, due to Hartree that computer applications in Cambridge got off to a good start.

A glowing tribute indeed.

As work on the EDSAC proceeded, Hartree foresaw a need for lectures on numerical methods. Over a period of several years, beginning already in 1947-48, Hartree presented a course of lectures, that he called "numerical analysis." In 1952, the lectures were published by the Clarendon Press under the same title. As the lectures, the book was intended to be introductory in the sense that no previous computational experience was assumed, but it was not deemed to be "elementary" in the sense of mathematical knowledge.

The introduction deals with the question "What is numerical analysis about?" The answer is that it is a *process*:

> Although from the point of view of numerical analysis the end to be attained is always a numerical result or set of results, the subject is not concerned with *results* ... but with the *process* by which the results can be obtained. And although the end is a numerical result, algebra and analysis are involved in the development of these processes. In so far as these processes, and the arguments by which they are derived, are general and independent of the particular values or numbers to which they may be applied, the subject may be regarded as a branch of mathematics. But the algebra and analysis must be aimed at providing or establishing *practical methods of obtaining numerical results*; otherwise it may be elegant mathematics, but not a contribution to numerical analysis.

He then went on to explain why the emphasis on *practical numerical processes* required a considerable change in attitude. One of his examples again was the solution of linear equations where the prevailing textbook method used determinants. In this he was correct. In the ensuing 50 years, algorithms for problems in linear algebra have been researched intensively and now are one of the best understood problems in numerical computation with reliable, general purpose software available.

Of historical interest, is his discussion of *Errors, mistakes, and checking*. He believed strongly in "checking" all results. He listed three ways in which the final answer might not be correct – truncation errors, rounding errors, and mistakes – and followed the list with the advice:

> Anyone intending to undertake a serious piece of calculation should realize that adequate checking against *mistakes* is an essential part of any satisfactory numerical process. No one, and no machine, is infallible, and it may fairly be said that the ideal to aim at is not to avoid mistakes entirely, but to find all mistakes that *are* made, and so to free the work from any *unidentified* mistakes. This of course is an ideal.

He considered carefully designed checking procedures to be important. Machines at that time were designed with thousands of vacuum tubes and acoustical delay lines for storage, and mistakes were made not only by humans but also by computers. The methods for checking that Hartree described were more suitable for hand calculations, but certainly, in the mid-1950's, it was not unusual to look at the second differences of a function to make sure they were smooth and did not exhibit the pattern of a mistake.

Hartree's book was a success. When reprinted in 1957 (also by the Clarendon Press at Oxford) he updated it to include many of the new results obtained from the use of computers.

The revised book ended with a description of methods for partial differential equations. This was an area of special interest to Hartree at the time, particularly equations arising in fluid dynamics. Hartree supervised two PhD students in this area who did their computations on the EDSAC. John Allen Ovenstone, from University of Sydney, did doctoral work on the mathematics of viscous hydrodynamics and boundary layer separation. He was awarded his PhD in 1953. A year later, Donald C F Leigh from Canada obtained his PhD, repeating the boundary-layer problem Hartree had laboured so hard to solve on a desk calculator prior to the war.

At the time of his death, Hartree was about to give a course of lectures on partial differential equations.

Teaching and the curriculum

Hartree had strong feelings about the mathematics curriculum in the Natural Science Tripos. As an undergraduate, he had spent a year studying for the Mathematics Tripos and then, upon returning after World War I, switched to the Natural Sciences Tripos since his interests were in the application of mathematics to the sciences, not pure mathematics. With the arrival of quantum theory, there were times when he considered his mathematical background inadequate for the task at hand. In a letter to Fritz London (September 16, 1928) in Berlin, written while Hartree was in Copenhagen at the Bohr Institute, he said:

> I'm afraid that having studied Physics, not Mathematics, when I was a student, I find group-theory very unfamiliar, and do not feel I understand properly what people are doing when they use it. [In England, "Physics" usually means "Experimental Physics;" until the last few years "Theoretical Physics" has hardly been recognized as a subject, like it is here, and, I understand, in your country. In Cambridge particularly, the bias has been very much towards the experimental side, and most people now

doing research on theoretical physics studied mathematics, not physics].

I have been waiting to see if the applications of group-theory are going to remain of importance, or whether they will be superseded, before trying to learn some of the theory, as I do not want to find it is going to be of no value as soon as I begin to understand something about it! Is it really going to be necessary for the physicist and chemist of the future to know group-theory? I am beginning to think it may be!

Now, as Plummer Professor in the Mathematics Faculty with connections to the Natural Sciences, he was in a strong position for urging the improvement of the courses in mathematics for those taking the Natural Sciences Tripos, in which several subjects could be studied, though mathematics was only a "half-subject" (one year program). He did not think the courses were meeting the needs of science students at all well. He brought about the promotion of mathematics to a "whole subject" (two year program) and took great interest in the design of the courses. New regulations went into effect for the year 1951-52. He undertook the development and presentation of the Part I Natural Sciences two year course in mathematics as a way towards obtaining the content and quality he believed was needed. Such courses were not the normal fare of a Professor but he was more concerned about undergraduate course quality than his own standing. Teaching mattered to him as much as research. His son, Richard Hartree, once asked him how much time he used to prepare his lectures. Hartree replied that with fifty or more students attending it was not unreasonable to spend a good part of fifty hours on one lecture!

Hartree was an extremely clear lecturer and took great trouble in making duplicate notes of his lectures available. This was before the days of xerox. He wrote them out by hand, quickly and legibly. When Léon Rosenfeld began teaching some of the courses at Manchester that Hartree had been teaching, he asked Hartree for a copy of his course notes.

Hartree also strongly advocated the modernization of the Mathematical Tripos, including Applied Mathematics, but the Department of Applied Mathematics and Theoretical Physics (DAMTP) was not established until a year after his death.

Notes and References

[1] Wilkes M V, Wheeler D J, and Gill S 1950 *The Preparation of Programs for an Electronic Computer* (Cambridge, MA: Addison-Wesley Press)
[2] Wilkes M V 1985 *Memoirs of a Computer Pioneer* (Cambridge: MIT Press)

Chapter 17

A trip to Australia

Hartree had many acquaintances in Australia. One was David Milton Myers (1911-1999), who had collaborated with Hartree and Porter on the use of the differential analyzer for the study of space-charge distribution in a triode (see Chapter 8). Hartree's enthusiasm for mechanical devices must have been contagious. Returning to Sydney, Australia, Myers joined the Radio Research Board of the Council for Scientific and Industrial Research (CSIR) as a research officer, having obtained a D Eng Sc (1936) for work including *Some mechanical aids to calculation* [1]. There he took up the design of an integraph for the solution of differential equations of the second order [2] which, unlike the differential analyzer that relied on shaft rotations, was designed around linear displacements. In 1939 he became Chief of CSIR's Division of Electrotechnology (ET). During WW II, he was involved in the development of accurate prediction methods for coastal defense gunnery and for the protection of shipping from magnetic mines. The technique consisted of de-Gaussing and was enormously successful.

Australia's need for the development and manufacture of radar systems and the production of microwave vacuum tube technology during the war, led to the establishment of a Division of Radio Physics (RP) within CSIR, headed initially by E G Bowen, one of the radar pioneers in the UK (see Chapter 12). By the end of the war the Division was in an excellent position to apply its expertise in radar pulse technology towards electronic computing. During the period 1945 to 1951, the transition to the computer occurred.

Early in this period it became clear to Myers, among others, that massive amounts of computing would be required by the sciences in the future. A Section of Mathematical Instruments (SMI) was established in ET as well as a Section of Applied Mathematics for research in mathematical methods, and for providing computational assistance and advice as needed. Later, in 1948, when Myers was appointed Chair, Electrical Engineering Department, Sydney University, the SMI moved with him. They developed a 10-integrator differential analyzer with electrical stepwise transmission of shaft rotations. At CSIR's Aeronautical Research Laboratory (ARL), Melbourne, a system of digital calculators were built.

Radio Physics took a different tack. Trevor Pearcey, born in the United Kingdom, had spent 1940-45 in the UK engaged in the mathematical aspects of short wave and microwave radar which required large-scale computation, using both analogue and digitals aids. He had worked with Hartree during the war on radio wave propagation, and had discussed with him the possibility of fast, digital computation. In late 1945, he joined the Radio Physics Division. There punched card machines had been installed for performing scientific computations and these were modified by Pearcey. In 1948, he along with Maston Beard, started the design of a stored program electronic computer, known as the Mark I. It was a "programmer's machine" in that it was structured not only for engineering simplicity but also economy and flexibility in programming. The Mark I ran its first program in November, 1949 and was believed to be the fourth stored program computer in the world and the first outside Britain and the US.

By late 1950, a number of activities were taking place, and advice was deemed prudent on how best to proceed. Hartree was contacted informally about a possible visit and must have conveyed his receptiveness to what was now called the Commonwealth Scientific and Industrial Research Organization (CSIRO). The Chief Executive Officer at the time was F W G White.

Frederick W G White (1905-1994) was educated in New Zealand. Upon graduation he went to St John's College, Cambridge, in 1929 to work with Rutherford. In 1931 he became interested in the amplitude of ionospheric reflections and went to King's College London where Appleton was Wheatstone Professor at the time. In 1936 he applied successfully for a chair of Physics in Canterbury University, New Zealand. In January, 1941, the Australian government requested White's services for three months in connection with radar but he never again returned to his position in New Zealand. From this brief outline of Frederick White's early career (he eventually became Chairman of CSIRO), the "village " aspect of physics during that period becomes evident again. Thus he and Hartree had many interests in common and, in fact, were well acquainted.

In letters to Bowen and Myers (October, 20 1950), White asked for suggestions on how to work out a logical case for the visit, a case to be put to the Executive Committee. Briefly,

(1) CSIRO has been developing machines and introducing the use of others
(2) We have machines which we can use for our work, but need to know:
  (a) is it worth continuing such developmental research?
  (b) would it be wise to stimulate wider interest in the use of such machines and in what direction is this likely to prove profitable?
(3) Hartree is the one man who has wide experience and who knows the position in the UK and USA. His advice would be invaluable.

Bowen responded that an important question to be asked near the end of the visit would be whether the work should continue in a number of centers as at present or whether it should be consolidated.

The Executive Committee approved a visit of six weeks during the period May 1 – September 1, 1951.

After brief discussions with Myers and Pearcey, the idea arose of organizing a conference in Sydney for the purpose of bringing together potential users and owners of computing machinery under stimulating conditions.

Thus matters fell into place and an official letter was sent to Hartree on January 2, 1951. He was to be reimbursed for return airfare, receive a subsistence allowance for six weeks, and a small honorarium. These were estimated later to be:

Air fare to and from Australia	£585
Fares in Australia	30
Expenses (42 days at £2.10.0 p.d.)	105
Fee	200

Hartree replied promptly. The period of six weeks seemed a bit too long for the program outlined. He suggested two possible dates and everyone agreed the later one was best in that it gave Pearcey the most time to get his machine ready. But White had also written privately to Hartree, mentioning how much he looked forward to the visit. Hartree confided to him that he was anxious to have his wife accompany him, but with such a small fee he did not see how he could afford her air fare. It was agreed that, if Hartree wished to present lectures at other Institutions there was no reason why he could not accept additional fees. Oliphant, then Director of the Research School of Physical Science, Australia National University was contacted, and immediately offered Hartree the needed financial support.

Marcus Laurence Elwin Oliphant (1901-2000) was born in Adelaide and went to Cambridge in 1927 where he also worked with Rutherford on nuclear physics experiments. As mentioned in Chapter 12, two of his team, Randall and Boot, invented the cavity magnetron that was crucial in the development of radar.

Oliphant's offer made all the difference. It was now possible for Hartree to bring his wife with him and he gladly fell in with the suggestion of visiting the University of Sydney, Melbourne, Adelaide, and Tasmania. Arrangements were made to leave Vancouver, Canada on Monday, June 25 arriving in Sydney on Wednesday, June 27 on BCPA (British Commonwealth Pacific Airlines) and leaving Sydney for Vancouver on August 11, 1951. He also planned to visit a computer project in Toronto, Canada for a few days either on the way out or the return.

Arrangements were made to hold a conference on automatic computing at the University of Sydney on August 7-9, 1951, sponsored jointly by the Electrical Engineering Department and CSIRO. The conference was organized into two sessions. The first session was general in nature, intended for those whose concern was the industrial, commercial or scientific application of computing devices. The second session was for those more directly involved in the design/development/operation of computing devices and also the programming of these machines.

Hartree played a major role in this conference as can be seen from the program, shown in Table 17.1. Of the 13 papers presented, he presented four whereas Myers presented three with collaborators.

The conference started with a welcoming address by Emeritus Professor, John Percival Vissing Madsen (1879-1969) who had been Professor of Electrical Engineering, University of Sydney, until 1949. He described the situation eloquently:

> Computing has traditionally occupied an intermediate position between theory and practice, and being neither fish nor fowl, has been looked at rather askance in the past by both species. But modern science and technology, with their rapid expansion accelerated by the demands of two world wars, have placed a correspondingly increasing requirement on the computer; consequently, both the producer and the user of mathematical theory have closed their ranks in stimulating the progress of computing methods and techniques.

The talks by Hartree were general in nature and some drew heavily on talks given at the University of Illinois in 1949. But much experience had

Table 17.1 Papers presented at the *First Conference on Automatic Computing Machines* held in August, 1951, in Sydney, Australia.

Welcome	Emeritus Professor Sir John Madsen, Kt, DSc, BE
Paper I	*Introduction to Automatic Calculating Machines*
	Professor D R Hartree, FRS
Paper II	*The CSIRO Differential Analyzer*
	Professor D M Myers and Mr W R Blunden
Paper III	*Automatic Digital Calculating Machines*
	Professor D R Hartree, FRS
Paper IV	*Digital Calculating Machines used by CSIRO*
	Messrs T Pearcey and M Beard
Paper V	*Introduction to Programming*
	Professor D R Hartree, FRS
Paper VI	*Programming for the CSIRO Digital Machine*
	Mr T Pearcey
Paper VII	*Automatic Calculating Machines and Numerical Methods*
	Professor D R Hartree, FRS
Paper VIII	*Programming for Punched Card Machines*
	Mr T Pearcey
Paper IX	*The Functional Design of an Automatic Computer*
	Mr T Pearcey
Paper X	*Some New Developments in Equipment for High-Speed Digital Machines* Professor D M Myers and
	Messrs D L Hollway, C B Speedy, and B F Cooper
Paper XI	*Some Analogue Computing Devices*
	Professor D M Myers
Paper XII	*Digital-Analogue Conversions*
	Mr W R Blunden
Paper XIII	*An Analogue Computer to Solve Polynomial Equations with Real Coefficients* Professor E O Willoughby
	and Messrs G A Rose and W G Forte

been gained at Cambridge in the use of the EDSAC during the intervening years and Hartree incorporated ideas from the book by Wilkes, Wheeler and Gill [3] to illustrate his talks, particularly in Papers II and V. A most important advance was the idea of subroutines which could be copied onto paper tape from a library of routines.

The papers were organized into sessions followed by lively discussions. Some related to the trade-offs between engineering simplicity and programming ease. The Cambridge policy had been to aim at simplicity of engineering whereas the CSIRO machine had a design that led to shorter programs.

One point came up several times and that was the use of the extra bits for hardware error-detecting codes. Hartree considered the use of error-detecting procedures at each operation of the machine as "counsel of despair." He claimed that experience had shown computers do not go wrong often enough to warrant this, and later mentioned that "the EDSAC has operated for hours without electrical of mechanical breakdown." Today, hardware error-correcting codes are standard.

Associated with the meeting was an exhibit of computing equipment, some of it loaned by various manufacturers and distributors of commercial type calculating machines, as well as the CSIRO Mark I and the Hollerith punch card machine.

While in Australia Hartree became ill and was diagnosed as being diabetic. It was not so severe as to need insulin treatment and he controlled it with diet. The Hartrees also took the opportunity to visit Rutherford's widow in New Zealand by flying boat from Australia.

In his earlier plans, Hartree intended to return to Vancouver, Canada but, possibly due to his illness, Douglas and Elaine took the opportunity of going on a trip around the world. They had traveled from London to Vancouver, Canada, then on to Hawaii and Australia. Instead of returning to Canada, they continued around the world. This was the era of piston-engined planes and it took them five days to return from Sydney to London, with overnight stays on the ground.

Notes and References

[1] Myers D M 1936 *Institute of Engineers, Australia, Journal* **8** 423
[2] Myers D M 1939 *Journal of Scientific Instruments* **16** 209
[3] Wilkes M V, Wheeler D J, and Gill S 1950 *The Preparation of Programs for an Electronic Computer* (Cambridge, MA: Addison-Wesley Press)

Chapter 18

Atomic structure research using EDSAC

At the University of Cambridge, Hartree supervised a number of students in atomic structure work. The first two performed hand calculations needed for studies of the solar corona and planetary nebulae, respectively. M Tuberg Gold spent 1948-49 at Girton College where she was supervised jointly by Bertha Swirles Jeffreys and Hartree. She computed radial functions, without exchange, for the $3s^23p$ and $3s3p^2$ configurations for Fe^{+13}. Roy H Garstang, Hartree's PhD student during 1949-1953, was the last to perform hand calculations, but including exchange. The ion he considered was Ne^{+2}.

Beatrice "Trixie" Worsley, who at one time had built her own Meccano differential analyzer [1], was a member of the Computation Centre of the University of Toronto where attempts were being made to build a University of Toronto Electronic Computer (UTEC). In the Fall of 1948, she was sent to Cambridge to learn everything she could. There she took part in the preparation for EDSAC's first run on May 6, 1949. At the Mathematical Laboratory she worked with both Hartree and Wilkes, the latter being her supervisor. Her dissertation on *Serial Programming for real and idealized digital calculating machines* considered three types of computations of which one was an automatic rendition of Hartree's SCF calculations with exchange. The dissertation contained a program written for the EDSAC with some preliminary results for Ne which M J Seaton used for photoionization calculations. These were the first numerical SCF calculations on the EDSAC.

The first published results to be carried out largely on the EDSAC were for gold and thallium, without exchange by S Douglas, Hartree, and Runciman. The calculations for thallium were undertaken as a refinement of the work on the theory of luminescence of solids where prior calculations had used the less accurate Thomas-Fermi approximation. Calculations for gold were, in effect, a repeat of work performed earlier by W Hartree, but with radial functions tabulated in a more convenient manner. Most of the paper consisted of a tabulation of the radial wave functions, typical of many papers of that era. The calculations on the EDSAC were performed by Hartree's student, Alexander Shafto Douglas, known as "Sandy." SCF calculations needed initial estimates of the field and it was helpful to use values derived from neighboring atoms or ions. Hartree's contribution to this paper was the description of the scaling of the contributions from the different groups of electrons so as to start with the best possible estimate of the field.

The next publication by Sandy Douglas [2] explored some new physics. It had always been known that, when an electron is placed outside a spherically symmetric charge distribution, the additional electron would distort this distribution. From classical considerations it was known that this "core-polarization" would modify the potential for the outer electron by a term that behaved as $1/r^4$, but what was its form closer to the nucleus? Some ideas were explored, yielding energy levels for the outer series electron in excellent agreement with observation for the times.

Cicely Ridley began her PhD studies with Hartree in Fall of 1953. Using the methods Hartree had outlined, but also including an idea that arose from a discussion with Sandy Douglas for automatically adjusting the energy parameter of the radial equation, she performed calculations without exchange for several heavy atoms — Mo^- [3] and In^{+3} and Sb^{+3} [4]. Five iterations at most were needed to adjust the energy parameter so that both the boundary conditions at the origin and infinity were satisfied. Another innovation was the use of the Runge-Kutta integration routine from the EDSAC library of subroutines implemented by Gill [5] for solving the second-order differential equation as a pair of first-order equations. This method would hardly have been feasible for hand-calculations but the use of a library routine reduced the labour and time for getting a working program.

In the Fall of 1954, Hartree acquired two new graduate students. David Francis Mayers arrived first in that he had worked at the Maths Lab during the summer. To David, Hartree assigned the problem of the relativistic Hartree equations, specifically mercury (Hg). This was the case Hartree had mentioned to Slater as an ultimate challenge, that he wished to see computed with Dirac radial wave functions *and* exchange, but he prudently decided that the first step was to perform the calculations without exchange. Some twenty-two (22) functions needed to be determined. The non-relativistic case was one Hartree had started on the Bush differential analyzer and finished with his father. No doubt he thought it would be interesting to compare the non-relativistic and relativistic wave functions. The other student was myself (then Charlotte Froese), recently arrived from Canada, and to me he assigned the problem of self-consistent field equations with exchange. Of course he was exceedingly busy. He made sure his students had the necessary information about the EDSAC and notes on atomic structure calculations, and over the Christmas vacation left for a year in the United States. His wish for an extended stay finally became possible.

During the previous years, Hartree had received a number of inquiries for results of calculations and for information about methods. Thus there seemed to be a revival of interest in the subject. When he was invited by Louis C Green to give a series of lectures as a Philips Visitor to Haverford College in the spring semester of 1955, he accepted. Much of the lecture material he presented dated from before the war, except for the rather condensed form of his review article with very little in the way of discussion of computational methods. Thus it seemed appropriate to collect all the material and provide a more complete description. In addition, the rapid development after the war of automatic digital computers, promised to make practicable many new methods. The Haverford lectures were repeated at the Palmer Physical Laboratory at Princeton University where Hartree was a Higgins Visiting Professor, possibly at the invitation of A G Shenstone. These lectures were the basis of his book on *Calculation of Atomic Structures* which he nearly completed while abroad.

David Mayers and I awaited his return to Cambridge with anticipation. During his absence there had been only modest communication. It was time for our results to be evaluated.

I had been developing a program, with exchange, for 10 and 12 electron systems and had applied it to F^- (required for some X-ray scattering work), Fe^{+16} and Fe^{+14} (extensions to high degree of ionization of theoretical interest to Hartree), and Al^{+3} and Al^+ (a check on some earlier calculations). Hartree always had a reason for every calculation! Initial estimates had been provided by Hartree and calculations for 12-electron systems were done under the assumption that the additional two electrons $(3s^2)$ did not change the core. An automatic energy adjustment process was devised, as had been done previously by C Ridley, and the Runge-Kutta-Gill library routine was used for the integration of the differential equations. With

Hartree's help, a paper was submitted on May 11, 1956 [6]. A conclusion of the paper was that more storage than the 1024 short words (16 bits plus sign) then available was required for effective calculations with exchange for more groups. Even so, a considerable amount of output to and from paper tape as auxiliary storage was needed. These were the first published calculations with exchange performed on the EDSAC, using Hartree's numerical methods.

Hartree also suggested that I perform the Ne^{+3} and Ne^{+4} calculations that had been requested by M J Seaton for the interpretation of spectra of gaseous nebulae, particularly planetary nebulae. The calculations were essentially complete before I went on vacation in August, 1956. Because of ideas of his own he wished to include, Hartree wrote up the paper containing the results during my absence, adding his name as second author. But he also described how he obtained initial estimates, stating "the better these estimates, the shorter the computation process."

The topic Hartree had been investigating most carefully in the mid 1950's was a means of getting the best possible starting values for calculations without or with exchange. For the former the starting point was a field which varied smoothly from the nuclear charge at the origin to the degree of ionization at large distances. For the "Fock equations" as Hartree sometimes called them, initial estimates were needed for each of the radial wave functions. How could the functions for one nuclear charge or configuration be interpolated/extrapolated to another in the same configuration or even a different configuration? Hartree proposed the following argument.

As the nuclear charge N increases and the number of electrons remains the same, intuitively, all electrons will be attracted more strongly towards the nucleus, but not necessarily by the same amount and rearrangement can occur. When only one electron is present and the wave function is said to be "hydrogenic" it can be shown that

$$N = \bar{r}^{(H)}(1)/\bar{r}^{(H)}(N) \tag{18.1}$$

where $\bar{r}^{(H)}(N)$ is the mean radius of a hydrogenic function for charge N. But in a many-electron system, some groups of electrons screen the nucleus as seen by other groups. So Hartree defined an "effective nuclear charge" for each group through the relation

$$N - \sigma = \bar{r}^{(H)}(1)/\bar{r}^{(HF)}(N) \tag{18.2}$$

where the right-hand side is the ratio of the mean radius of the wave function for hydrogen and $\bar{r}^{(HF)}(N)$ the mean radius of a Hartree-Fock (or Hartree) radial function. This relation defined a screening parameter σ. By plotting σ as a function of nuclear charge for different configurations, he was able to estimate the screening parameter for a new configuration and get reasonable initial estimates for an SCF calculation. Because of the rearrangement, it was desirable to scale each radial function individually, even for calculations without exchange.

Hartree had a talent for understanding how quantities varied so that, through transformations, others could be defined whose variation was almost linear. In the case of radial wave functions, he introduced the "reduced radius," $s = r/\bar{r}$, and the "reduced wave function,"

$$P^*(nl, s) = \bar{r}^{1/2} P(nl, \bar{r}s)$$

and thought of these as functions of the mean radius. In our paper, he included Figure 18.1 that showed the systematic variation of a $2p$ reduced wave function for fixed groups (solid line) and for fixed N (dotted line) each as a function of the mean radius. Indeed, the variation is remarkably linear and from such tables, reliable initial estimates could be obtained. The calculations, on which this figure was based, had been performed by Hartree using a program drawn up by J M Watt, a program that computed mean radii and reduced wave functions and involved a considerable amount of non-linear interpolation.

Hartree was not a coauthor of many of the papers published by his students. He believed that "If you did the work and got it right, you deserved the praise and if it was wrong, you deserved the blame." Good examples are the papers where Hartree provided initial estimates and general guidance: in those cases he was not included as a coauthor. Instead, he was the person refereeing the paper and submitting the manuscript, generally to the *Proceedings of the Cambridge Philosophical Society*.

It should be mentioned that computing on the EDSAC, the world's first general purpose computer, was quite unlike computing today. The machine was built out of vacuum tubes which ultimately would fail. In such a case, the engineers needed to find the defective tube and replace it, which could be a tedious process. Secondly, memory consisted of mercury delay lines. Pulses were generated at one end of a tube filled with mercury, picked up, regenerated at the other and recycled. The velocity of a pulse in a column of mercury depends on the temperature and, as the latter varied throughout the day, adjustments were needed to keep the memory

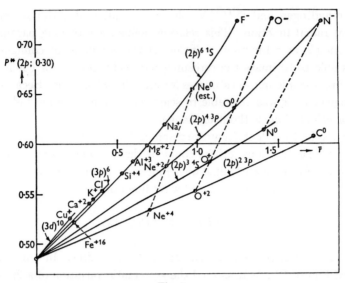

Fig. 1

Fig. 18.1 Variation of a reduced wave function $P^*(nl; 0.30)$ as a function of the mean radius for the $2p^6$, $2p^4$, and $2p^2$ isoelectronic sequences (solid line) and these configurations for $N = 7, 8, 10$ (dotted lines) as a function of the mean radius.

functioning properly. The EDSAC was on the top floor of what had been an anatomy building and there could be considerable temperature variation during the course of a 24-hour day. There was no environment control other than a door to the roof of an adjacent building which could be opened or closed. The proper functioning of the machine required periodic tests and machine adjustments. During the day, the engineers were usually either performing maintenance or making improvements to the machine, and staff were performing special runs. At tea-time, morning and afternoon, users could line up with their paper tapes containing the machine instructions to be executed, and perform short test runs. Research production was done at night, usually starting at 6:00 PM when the engineers left to go home, until the machine was no longer functioning properly or the first engineer arrived in the morning.

At the beginning of every term, all the approved users would meet and "nights" were assigned [7]. Faculty with their diploma students usually got the early hours. David Mayers and I along with a few others, were usually assigned to Monday evenings with J C P Miller, a number theorist, and his students getting the 6:00 PM shift. The frustrating fact was that number theory could be done using only addition and so, if there were problems with the multiplier, say, the number theory programs would execute without complaint. There always was a chance that when David and I got to run our programs, the machine might be malfunctioning. Many a night was spent working with the machine. Going home meant riding a bicycle in the dark, wearing a Cambridge gown which I was required to wear after dark, while the University was in session. But every once in a while, things went well and one or the other of us would work through until the morning.

"Trixie" Worsley had returned to the University of Toronto, Canada at the end of the summer of 1950 where she was on the staff of the Computation Centre. In April, 1952, the University installed a Ferranti machine, which Trixie had suggested be named "FERUT" after "FERranti University of Toronto." There she completed her work on a program for atomic structure calculations, directed by Hartree who continued to act as a consultant. In 1957 she submitted a manuscript for publication [8] describing a FERUT SCF program and presenting results with exchange for neon. She had gone back again to the $\log r$ variable Hartree had used on the differential analyzer but now for a quite different reason: in an SCF calculation with exchange, much time is spent computing certain functions that could be computed most rapidly in the $\log r$ variable using an integrating factor. But it also is of historical interest in that the paper mentioned a set of programs that had been developed for electronic computers at that time [9]:

(1) B H Worsley, 1952 for the EDSAC
(2) A S Douglas, 1955 for the EDSAC
(3) E C Ridley, 1955 for the EDSAC
(4) C Froese, 1957 for the EDSAC
(5) S L Altmann, 1955 for Manchester University Computer Mark II
(6) W W Piper, 1956 for the IBM 650 Computer
(7) B H Worsley, 1958 for FERUT

But the programs could not really be compared because, as she said:

> Considerable variety has been shown in these programs in many particulars. These include such aspects as the extent of applicability, the amount of judgment coded into the program, the formulation of the equations for evaluating the potential and exchange terms as well as of the Hartree-Fock equation, the choice of independent variable, the selection of the intervals of integration, the technique for satisfying the conditions of orthonormality simultaneously with the boundary conditions

in solving the wave equation, and finally such details as the numerical representation, scaling, and decimal-point location.

The last remark relates to the fact that early computers did not support floating point arithmetic and it was the responsibility of the programmer to scale numerical values as required by the particular computer being used. In preparing this list, Worsley included only those programs that followed Hartree's basic numerical procedures. Not included was work by S F Boys and his group using a non-orthogonal set of analytic functions. The earliest paper was by Boys for the beryllium ground state [10] and appeared in 1950. In his conclusion, Boys admitted that "the theoretical part of the calculations is more complicated ... but the numerical work is apparently considerably less."

Since those early days, the Hartree-Fock method has expanded into new atomic physics areas – time-dependent collision processes, atoms in strong magnetic fields, material science, nuclear theory, and most recently in Bose-Einstein condensates. Hartree-Fock theory together with "superposition of configurations" is also at the core of quantum chemical calculations.

Notes and References

[1] Williams M R 1994 *IEEE Annals of the History of Computing* **16** No 2, 4
[2] Douglas A S 1956 *Proc. Camb. Phil. Soc.* **52** 687
[3] Ridley E C 1955 *Proc. Camb. Phil. Soc.* **51** 702
[4] Ridley E C 1956 *Proc. Camb. Phil. Soc.* **52** 698
[5] Gill S 1951 *Proc. Camb. Phil. Soc.* **47** 96
[6] Froese C 1957 *Proc. Camb. Phil. Soc.* **53** 206
[7] The EDSAC, or EDSAC 1 as it was later called, became operational in May 1949 and was finally switched off early in 1958. Thus during my stay at Cambridge, from September 1954 to September 1956, EDSAC 1 was in the process of being replaced by a much easier to use EDSAC 2.
[8] Worsley B H 1958 *Can. J. Phys.* **36** 289
[9] The dates mentioned refer to dates of publication of a paper announcing the work. The actual program development may have been completed a year or so earlier.
[10] Boys S F 1950 *Proc. Roy. Soc. A* **201** 125

Chapter 19

The final years

In the Fall of 1956, I returned to Canada, David Mayers left for the Atomic Weapons Research Establishment at Aldermaston and no new students arrived on the scene. I was trying to complete my dissertation within a year. The Cambridge degree requirements included only two years in residence, after which arrangements could be made for completion of the degree at another institution. In my case, Dr W H Watson, then chairman of the Physics Department at the University of Toronto, was appointed as my supervisor. Before my departure from Cambridge, Hartree mentioned that a dissertation should not only include computational results, but also some theory. Here he proposed some "Z-dependent" theory, as it is now called, where Z is the nuclear charge. About half the resulting dissertation consisted of tables of numbers.

In the coming year, Hartree and I communicated fairly frequently – there were papers submitted that needed to be monitored through the publication process and then there was the dissertation itself. Figure 19.1 shows parts of a letter with his very distinctive signature. The letter also shows his desire to be informed about the work of others since he did not want to compete in any way.

For a dissertation to be accepted, it had to be evaluated. If it was not possible for me to return to Cambridge for an examination, a written examination could be taken. Fortunately, Hartree was invited by Slater to present a paper at a conference at his Solid State and Molecular Theory group at MIT and to stay on as consultant for a few more weeks. Hartree generously suggested that he come to the University of Toronto for the thesis defense which simplified matters immensely. During the preparation of the thesis he was extremely helpful communicating on numerous occasions with suggestions. The examination itself was an informal meeting of myself

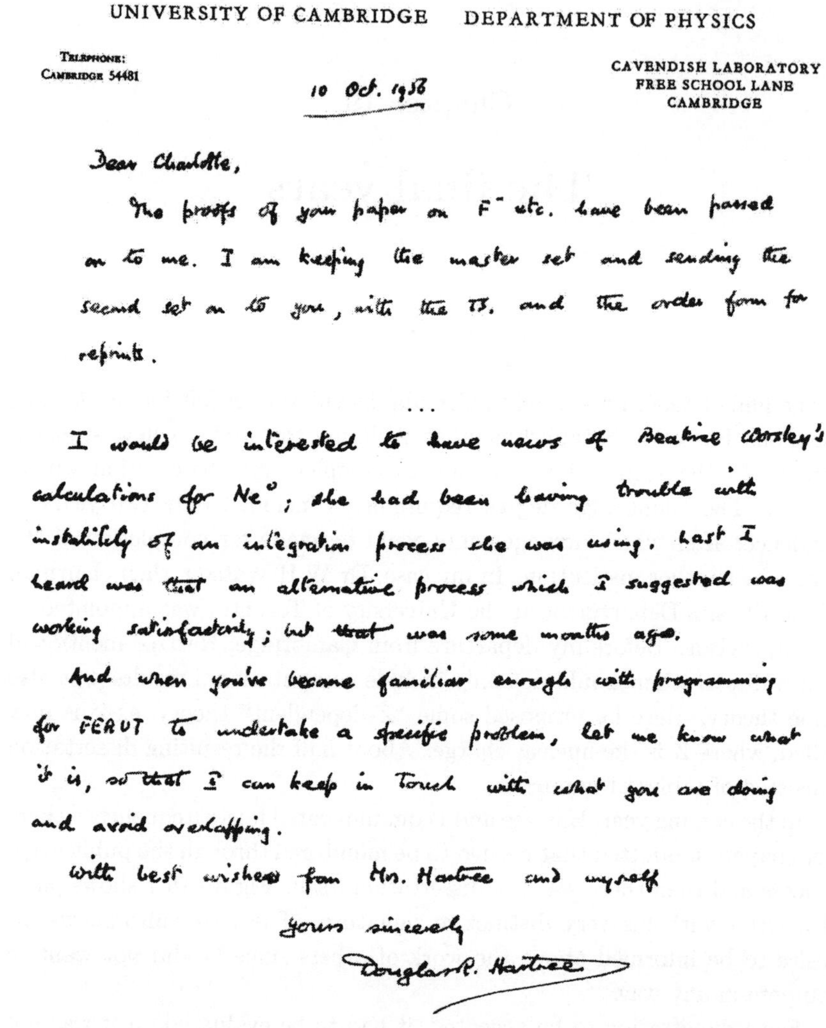

> UNIVERSITY OF CAMBRIDGE   DEPARTMENT OF PHYSICS
>
> TELEPHONE:
> CAMBRIDGE 54481
>
> 10 Oct. 1956
>
> CAVENDISH LABORATORY
> FREE SCHOOL LANE
> CAMBRIDGE
>
> Dear Charlotte,
>
> The proofs of your paper on F^- etc. have been passed on to me. I am keeping the master set and sending the second set on to you, with the TS, and the order form for reprints.
>
> ...
>
> I would be interested to have news of Beatrice Worsley's calculations for Ne^0; she had been having trouble with instability of an integration process she was using. Last I heard was that an alternative process which I suggested was working satisfactorily; but that was some months ago.
>
> And when you've become familiar enough with programming for FERUT to undertake a specific problem, let me know what it is, so that I can keep in touch with what you are doing and avoid overlapping.
>
> With best wishes from Mrs. Hartree and myself
>
> Yours sincerely
> Douglas R. Hartree

Fig. 19.1   Copy of part of a letter Hartree wrote on October 10, 1956.

and Professors Hartree and Watson. It was a friendly meeting, memorable for the fact that Watson asked "Do you believe in parity violation?", an idea that I had never heard of at the time. After floundering for an answer, Hartree himself asked Watson "What do you think?" At that point they discussed the matter for a while, much to my relief.

David Mayers too was making progress. In due course a paper presenting results of self-consistent field calculations for mercury was ready for submission. Hartree must have thought highly of this work for on February 20, 1957 it was submitted to the *Proceedings of the Royal Society*.

One of Hartree's last tasks in 1957 was to convene a meeting of the Mathematical Tables Committee. Mathematical tables had been valuable aids to computation for decades. It was the mechanization of table making that had first motivated Babbage in his design of the difference engine. In response to the importance of accurate tables, the British Association for the Advancement of Science, established a Mathematical Tables Committee in 1871. The list of Committee members included many outstanding mathematicians dedicated in their search for perfection in accuracy and typography in table making [1]. But with the development of computers, interest in tables declined. Charles Darwin was asked by the British Association Council for his advice and he concluded that funding for table making was no longer needed. At that point, the British Association asked the Royal Society to take over the responsibility and the transfer took place in June, 1948. The first Chairman of the Mathematical Tables Committee of the Royal Society was Darwin himself, with Hartree a member. This new Committee tended to act primarily in a co-ordinating role of bringing the community closer to computers. The lack of enthusiasm for tables was evident from the fact that only four tables were published during the period 1948-58. Hartree was elected Chairman of the Committee in 1957. His last meeting as Chairman was December 1957. By the next meeting on March 21, 1958, Hartree had died and Wilkes succeeded him. It is doubtful that Hartree achieved much in his one year as Chairman at this time of the Committee's history. He had not been a particularly active member during prior years. When Wilkes became Chairman, he saw his task as bringing closure to table making. In July 1965, the Royal Society dissolved the Committee, replacing it by a small advisory body.

Hartree was diabetic. Early in 1958, Douglas and Elaine became specially concerned about his health. His doctor recommended that he have some tests performed at a hospital. On February 12, 1958, on his way to the private hospital about a quarter mile from his home, he collapsed on the pavement. A passer-by called an ambulance but he had suffered a severe cerebral hemorrhage and was dead on arrival at Addenbrooks Hospital. He was 60 years old. That very same day, a granddaughter was born.

Obituaries appeared in *Varsity* (February 15, 1958), *Cambridge Daily News* (February 15, 1958) noting that he was the son of former Mayor, *Manchester Guardian* (February 14, 1958), *The Times* (February 14, 1958) followed by a personal note by Lady Jeffreys (February 21, 1958), *Journal of IEE* (April 1958) by Arthur Porter, and *Cambridge Review* (March 1, 1958) by Nevill F Mott. Two others were written by Maurice V Wilkes, one of which appeared in *The Computer Journal* of the newly formed British Computer Association, noting Hartree's comment that he had logged some 10,000 hours of computing during his lifetime, and another on March 22, 1958. The funeral services were held February 21, 1958 at Christ's College Chapel.

His last paper, published after his death, was a publication of radial charge densities for the Ti^{+2} core that had been sent to R J Weiss of the Ordnance Materials Research Office, Watertown Arsenal, Massachusetts, USA and was communicated to the Proceedings of the Cambridge Philosophical Society by W H Taylor, a family friend. Thus his research career ended very much as it started – with differential equations and atomic structure calculations. With respect to the latter, Slater said, comparing Hartree with the "magical" type of scientist [4]:

> Yet he made a much greater contribution to our knowledge of the behavior of real atoms than most of them did. And while he limited himself to atoms, his demonstration of the power of the self-consistent field for atoms is what has led to the development of the method for molecules and solids as well.

But in the intervening years he also was a strong advocate for, first, computing devices and, then, computing machines. From his range of computational experience, he distilled an understanding of the computational process, and the role of numerical analysis as a branch of mathematics. In today's world he would have been comfortable as a "Computational Scientist," working at the interface of mathematics, computing and applications.

Sadly, J Womersley died just a month later on March 7, 1958, at the age of 50 years. Charles Darwin wrote the Royal Society *Biographical Memoirs* for Hartree [2] and an obituary notice in *Nature* for Womersley [3]. Darwin himself died on December 31, 1962 at the age of 75 years.

An era had passed.

Notes and References

[1] Croarken M and Campbell-Kelly M 2000 *IEEE Annals of the History of Computing* **22** No 4, 44.

[2] Darwin G C 1958 *Biographical Memoirs of Fellows of the Royal Society* **4**

103
[3] Darwin G C 1958 Obituaries: J R Womersley *Nature* **181** 1240
[4] Slater J C 1975 *Solid State and Molecular Theory: A Scientific Biography* (New York: Wiley-Interscience)

Chapter 20
His legacy

During his entire scientific career, Hartree was devoted to ensuring the development and advancement of his students and assisting all those who came to him with scientific ideas or problems. Included here are comments about a few individuals who have recognized his contribution to their professional lives and remarks from some, but by no means all, of his students.

20.1 Bertha Swirles Jeffreys

Biographies of Bertha Swirles Jeffreys usually mention that she was R H Fowler's research student, but in the biographical memoirs of her husband, Sir Harold Jeffreys, it is mentioned that D R Hartree also was her supervisor. Probably because he had just completed his own PhD, he was able to pass on problems and provide guidance. A close professional relationship remained between them over the years. Both as a faculty member at Manchester University and later at Cambridge University, Bertha Swirles Jeffreys was a very special colleague. After Hartree's death, Bertha wrote several articles in memory of Douglas Hartree that have been an invaluable source of information for this biography.

Like Hartree, she had a love for music. She was a skilled musician, played the piano and cello, and was Director of Studies in Music at Girton College for eight years. She died in December, 1999 at the age of 96 years.

20.2 Arthur Porter

Hartree's first PhD student was Arthur Porter whereas I was the last to begin and complete my PhD under his direction. On Monday, February 8,

2000, I (then Charlotte Froese Fischer) interviewed Arthur Porter and his wife Patricia, at their home in Naples, Florida.

C: How did you happen to work with Hartree?

A: I was studying Physics at the University of Manchester and during the finals year a research project was required. Brentano, the supervisor of the laboratory asked "Porter, what are you really interested in?" I said "Wave mechanics!" He said in no way would it be possible. So then I said "Computers!" That came completely out of the blue. The same morning in browsing through a recent volume of the Cambridge Philosophical Society Proceedings, I had seen a paper by Edward Bullard on the use of a moving coil galvanometer to solve second-order linear differential equations and I was intrigued. Brentano said, "Well, I had lunch today with Professor Hartree who has just returned from America, and he asked me if any student in my class was interested in computers. You had better go over and see Professor Hartree right away."

C: So this was a project for your baccalaureate degree.

A: That's right, a very small part of the differential analyzer (DA) project.

C: And then you went on to do graduate work.

A: Correct - under Hartree's supervision. I was fortunate to obtain graduate scholarships for the period 1933-36 and my program related specifically to the construction and applications of the Meccano DA and subsequently of the full-scale analyzer with 8 integrators.

C: What did you consider some of your most significant achievements, or most interesting work?

A: I think the model DA constructed largely of Meccano was most interesting. Of course Douglas Hartree was an absolute authority on Meccano since he had three young children, and he introduced them to Meccano. After seeing the Bush DA at MIT his first impression was "Gee, this is just glorified Meccano." The machine performed far beyond our expectations. The wave functions of the chromium atom that it produced were smooth, without a tremor. The Meccano DA is now on display at the Science Museum in London.

The other field that attracted me was control theory. The first project was given to us by Albert Callender at ICI (Imperial Chemical Industries). The problem essentially was to determine the effect of finite time lags in control systems. These arise in chemical plants. The initiation of a valve has associated with it the time it takes for the effect of that action at a point some distance away. This is a time lag. Callender stated the problem fairly succinctly and he built an analogue model of the process, Hartree formulated the problem mathematically, and I obtained solutions of the control equations using the Meccano differential analyzer.

C: Did you get to know Vannevar Bush?

A: Oh, very well. He was not a distant person. His interest in young people was the catalyst of the whole operation. Really, two years of encouragement were quite rewarding. Hartree was exactly the same of course, so there I had the great good fortune to be associated with Douglas Hartree at Manchester and Vannevar Bush at MIT, as well as the other people in the group. This was now the Rockefeller DA, on a much larger scale than the first and conceptually quite different because it was analogue with some digital.

C: Nick Eyres, was he a Baccalaureate or PhD student?

A: I don't think he was a student. He came into the group because when the war started (WW II) it was very difficult to obtain qualified people to work the DA. Most of the scientific staff had gone into radar, ballistics, fire control, or computers directly associated with the services. So Douglas recruited Nick Eyres, the first cousin of Elaine Hartree.

Mrs Patricia Porter: Douglas Hartree lived with us at Hampton Hill for about 6 weeks just after the war. He was doing some work at the National Physical Laboratory, a mere two miles away, and Elaine asked if he could stay with us. He used to do all his work on his knee. He never sat at a desk. He had a sort of board that he carried around with his briefcase. My recollection is of him sitting around for hours writing and doing mathematics.

A: Or playing with our son with his electric trains. DRH was addicted to railways, especially signaling systems. His idea of a restful Saturday afternoon was to spend the afternoon in the main signal box at London Road Station, Manchester. On one occasion he invited me to join him and his two boys!

C: What other memories of Hartree do you have?

A: Hartree frequently used the slide-rule during his journey from Manchester to London. After one trip he apologized for not having completed a calculation, explaining that the train had been crowded and his elbow was bumped a lot. He loved music and must have spent hours operating his "player piano." He was a tympanist with the University Orchestra at Manchester. A constant memory of Douglas is the Hartree "stool," a prized possession which resides in my den. It was a wedding present from Elaine and Douglas (57+years ago). The stools were made by a furniture craftsman in Kendal and were given to several of his early students and colleagues. The only other recipient I know well is Sir Bernard Lovell (of Jodrell Bank fame).

20.3 Jack Howlett

The differential analyzer was probably Europe's most powerful calculating engine when World War II began. As a result of being recruited to Hartree's differential analyzer group during the war, a group that made valuable contributions to the atomic bomb project, Howlett had the opportunity of taking charge in 1948 of what became the Harwell Atomic Energy Research Establishment's computing section. In 1958 Harwell acquired a Ferranti Mercury computer and together with Manchester University and Ferranti planned the Atlas project. In 1961 Howlett became director of the Atlas Computer Laboratory.

Upon his death, his obituary in *The Guardian Unlimited Archive*, which appeared on Thursday May 20, 1999 started with the sentence:

> He helped to boot the computer out of the mechanical era and into the electronic one

Hartree left his mark on "Mathematician, Jack Howlett."

20.4 John Crank

John Crank studied Physics at Manchester University (1934-38) under Sir Lawrence Bragg and Professor D R Hartree and received an Hons BSc and MSc and in 1953 a DSc. On March 9, 2000 he wrote:

> As a physics undergraduate at Manchester I took a course in Applied Mathematics under DRH. He was very lucid and painstaking and, as far as I recall, never used any lecture notes. After graduation, I joined his research team and worked on his Differential Analyzer. In my MSc thesis (1938) I acknowledged his keen interest and assistance with the

thesis and the working on the differential analyzer. Nine years later in the preface to my book *The Differential Analyzer* (Longmans, Green and Co), published in 1947, I acknowledge my indebtedness to DRH who read the original manuscript and made many helpful suggestions. I see that my friends, Mr and Mrs M Nicolson are also thanked for reading the proofs.

I recall that DRH's hobby was Model Trains and he gave a talk about this in Dalton Hall, the Student Residence where I lived.

During the war years, I was in charge of the newly installed DA in the Cambridge Mathematical Laboratory. In the early 1940's DRH was asked by the Cambridge Physical Chemist, Dr C H Bamford, to look at the mathematical problem of burning wood, on which he was doing experimental work. I put the problem on the Cambridge DA and under DRH's direction, Phyllis Nicolson, one of Hartree's group in Manchester, tried a numerical approach which was giving her difficulties. Later we collaborated and produced the Crank-Nicolson method.

20.5 Nicholas Eyres

On November 24, 1998, Nick Eyres wrote:

I worked under Hartree's direction 1940 - 1945 in a small group operating his beloved differential analyzer (after Vannevar Bush at MIT in the 30's). This was then the only practical way of solving the many equations thrown up by wartime research establishments.

This vast field of computing was a totally new experience for me, having a few years earlier been through the mill of Cambridge maths, then very formal. It entirely changed my outlook as a teacher for the rest of my life. Exposure to this work in general and Douglas and his way of thinking, in particular, were the best things that ever happened to me professionally.

20.6 Sir Aaron Klug

Aaron Klug came to do his PhD in the Cavendish Laboratory in 1949 [1]. He had obtained his MS from Cape Town University under the directorship of R W James, an X-ray crystallographer who had been a colleague of both Lawrence Bragg and Hartree at the University of Manchester. At Cambridge, he wanted to continue the study of X-ray crystallography of some sort, such as protein structure, but after a few weeks, he decided to take a project that Hartree proposed, left over from the war. Hartree was doing heat transfer calculations on the cooling of steel ingots. They were made by quenching the hot steel quickly so that it passed through the austenite-pearlite transformation rapidly (which results in the hardening of steel),

Traditional practice was to leave an ingot for nine hours overnight, but Hartree did some calculations to show that the reaction was complete by about four or five hours, thus saving a good deal of time and speeding up production. In order to solve the differential equations he had set up to simulate the process, Hartree used an artificial thermal diffusion coefficient which was variable with temperature. He deduced its value from laboratory experiments on the cooling of small cylindrical bars of hot steel. This approach was sufficient to give a rough approximation of the cooling process, but there was no understanding of what was going on inside.

Hartree told Klug he would like someone to tackle the problem, introducing the right mechanism of transformation. Klug had done some X-ray crystallography and knew something about materials. Hartree thought a crystallographer was needed to understand the process and to calculate it properly.

After learning some metallurgy, Klug set up a scheme in which the new phase (pearlite) was nucleated in the austenite matrix and gradually grew. He developed a nucleation growth model, the rate of transformation being governed by the dissipation of the latent heat of transformation. He ended up using numerical methods to solve the partial differential equations for heat flow in the presence of a phase transition. He learned a good deal during this time, particularly in computing and solid state physics.

After obtaining his PhD, Hartree recommended him to Dr F J W Roughton in the Colloidal Science department who wanted someone to help him tackle simultaneous diffusion and chemical reaction, such as occurs when oxygen enters a red blood cell. The methods Klug had developed for steel were applicable here as well. When he worked on the assembly of virus particles out of their constituent protein units and nucleic acid, he again used a nucleation and growth model, but this was now in the field of physical biochemistry. In time he became more and more interested in biological matter and in 1982 was awarded the Nobel Prize in chemistry "for his development of crystallographic electron microscopy and his structural elucidation of biologically important nucleic acid-protein complex."

20.7   Roy H Garstang

On July 8, 1998, Roy Garstang wrote:

> My impression of Hartree is that he was very shy. After I got to know him a bit he was very friendly, always very kind, and willing to help as, for example, with some estimates of wave functions. I took his course on Numerical Analysis in my Mathematical Tripos Part III as one of my examinable courses, and I learned a lot from him. I was invited to tea at his house once. I still have a record of the date – it was February 19, 1950. That was my only social interaction with him or with his wife. This may have been because of shyness, or it may have been because he was always off to meetings in London and very busy. Lennard-Jones was the same in this respect. They did not keep office hours in the American manner. You had to catch them if you wanted to see them even for a few minutes. If he was in Cambridge, Hartree would sit in his little office with an open door, and he would always stop what he was doing and talk to you.
>
> Hartree went to the US for a one term sabbatical, probably the autumn of 1950, and I was supervised by Bertha Jeffreys. In addition, Hartree was away in the autumn of 1953, and my PhD examiners were David Bates from Belfast, and Bertha Jeffreys, replacing Hartree.
>
> One other thing that struck me about Hartree was the time he spent doing things for others, like solving equations etc. He was most unselfish with his time, and probably enjoyed all the contacts which he had.

20.8   Donald C F Leigh

In April, 2000, Donald Leigh wrote:

> In 1951 I graduated from the University of Toronto with an undergraduate degree in Engineering Physics. I applied for a fellowship at several universities including a new set offered by the United Kingdom to Canadian students in order to have Canadians become more familiar with British productivity. In my interview with the selection committee, somehow it was suggested that I talk with the director of a British research lab in Toronto. He happened to be one of Hartree's past graduate students and suggested that I apply through Hartree which I did. I was admitted to the graduate program with Hartree as my tentative advisor.
>
> My research was to develop a numerical solution of a fluid mechanics problem using a computer called the EDSAC: it was the first *internally programmable* computer that actually functioned. By contrast the American ENIAC was an *externally programmable* computer. Hartree became

my advisor and I very much enjoyed being his student. I received my PhD in 1954 and I published a paper the same year.

20.9 David F Mayers

In May, 2000, David Mayers wrote:

> In my third year of mathematics at Cambridge, in 1953, I chose, among other options, to study Quantum Theory with Dirac, and Numerical Analysis with J C P Miller; I can't remember why I chose these, but they were a good preparation for the future. The Numerical Analysis lectures were given in the lecture room on the top floor of the Mathematical Laboratory, so I got to know my way around the building. All the practical work was done on desk calculators, as undergraduates were not allowed too near EDSAC.
>
> About this time I answered an advertisement for somebody to spend some weeks of the vacation helping with low grade electrical work in the Maths Lab. During that summer I got to know several of the people there, and Maurice Wilkes passed me on to Hartree as a possible research student. Hartree managed to raise some funding, and I started the PhD work in 1954.
>
> A few days after I officially became a research student, Hartree gave me about four pages of notes and instructions, and went off to the USA for a year, so I learned programming from Wilkes, Wheeler and Gill, and of course, Sandy Douglas, and fellow students, including Charlotte Froese (the present author).
>
> Hartree had started me off with the relativistic form of the Hartree equations, applied to the mercury atom. Soon after, he also passed on to me a problem from Gerry Brown, who was then working with Peierls at Birmingham. This involved the solution of systems of similar differential equations, involving Bessel functions, and the evaluation of a number of integrals. Hartree suggested that the Bessel functions could be calculated by solving the Bessel equations along with the others, and the integrals could also be calculated by solving very simple differential equations. The problem then reduced to the solution of a system of 39 coupled differential equations. Hartree suggested that this was the best way to go about it, and I got on with it as a matter of course, though some other people around found it remarkable that such a large system could be solved numerically. Nearly 50 years later it looks a simple problem, but it still would not be trivial to solve on a computer with only 4K of memory; at the time Hartree had a very clear idea of what could be done with the computing power available.
>
> At the same time I was struggling with the wave functions for mercury. After finally completing one iteration round all of them I sent off

some numbers to Hartree in the USA. I didn't keep his reply, or anyway I can't find it, but I remember it now as a model of how to treat an inexperienced student. He could have said that the numbers were obviously rubbish - which was true; but he just commented that they were not quite what he had expected, and explained why. He also suggested a number of checks that I could make - and should have made before. Before long I discovered my appalling misunderstanding of the relativistic model, and once that was put right the next iteration was satisfactory. By the time Hartree was back in Cambridge the whole project was going pretty well. Also by this time Stella and I were married, and he and particularly Mrs Hartree were very kind to us as we settled in to our new home.

Like all the best Supervisors of research students, Hartree ran a very extensive and efficient Employment Agency. In 1956 he arranged for the offer of a junior research post at AWRE, Aldermaston, where the IBM 704 was just being installed. This involved going on with very much the same line of work, in Walter Stibbs' group, along with Dick Carson and Ian Grant. Ian picked up the relativistic HF equations, and has continued to run with them ever since. Then in 1957 I had a letter from Hartree, suggesting that I should apply for a job in the new Computing Lab at Oxford; no doubt he also said the right things to a number of the right people, as I got the job.

After his death we continued to keep in touch with Mrs Hartree; back in Cambridge for the PhD Degree ceremony, she took charge of Christopher, our new baby, so that the rest of the family could take part in the celebrations.

20.10 Ronald J Lomax

Dr R J Lomax, presents another facet of DRH:

I actually did my PhD work under the direction of Oscar Buneman (Peterhouse). However, he went to Stanford on sabbatical leave 1957-1958, so it was necessary for me to have someone appointed as advisor for the year in his absence. This is where Hartree came into the picture. During World War II, Hartree and Buneman worked together at the University of Manchester and were involved in work on magnetrons for military RADAR. Under Buneman, I had been doing work on electron-beam confinement (electron guns), and in general Buneman's interests at that time were in charge-particle flow including electron devices and plasma physics. While I don't think Hartree was doing any work in this area at the time, he was actively involved with numerical analysis and the early electronic computers. Anyway, Buneman asked Hartree if he would act for me in his absence, and Hartree agreed.

Hartree showed me some work he had done during the war, which involved calculations on the flow of electrons in vacuum tubes (valves in British parlance). He still had the results of these calculations, which were done by hand using mechanical calculating machines. He had published some of this in a paper *Some calculations of transients in an electronic valve*, Appl. Sci. Res. vol. B1, 1950, p. 379. He suggested that I take a look at repeating the calculations on EDSAC 2, and extending them considerably. It turned out that there was some very interesting oscillatory behavior that had not appeared in the hand calculations because the pattern had not had time to get established in the laborious hand calculations. From this and my work with Buneman, I became interested in electron devices and plasma simulation. The work I did on plasmas never came to much and subsequently I became interested in solid-state device simulation, and later in very large scale integrated circuits (VLSI).

As you know, Hartree died very unexpectedly early in 1958. At this point, I had to have yet another research supervisor - K F Sander from the Department of Engineering. Since I was still working on the ideas that Hartree had suggested, Sander did not have much impact on the direction of my work, and Buneman returned in fall 1958.

I finished my PhD in 1960. I had been appointed to a Bye-Fellowship at Peterhouse in 1959, which I held until 1961 when I came to the US with the intention of spending a year as a visiting assistant professor at The University of Michigan - and I am still here 38 years later in the Deptartment of Electrical Engineering and Computer Science, (although I am retiring next summer.) Buneman subsequently went to Stanford permanently as Professor of Electrical Engineering. I spent a year there on sabbatical in 1978-1979, but not working with him, although we did, of course, renew our acquaintance. Oscar Buneman died about four years ago, still at Stanford.

As you see from this, I was not working in the area that Hartree is traditionally known for, and unfortunately, it was only for a very short time. He was certainly a very friendly and encouraging person, and very unassuming despite his eminence. I am very glad that I had that brief opportunity to work with him.

20.11 A personal tribute from the author

Hartree's sudden death, just as I was starting my professional career, left me without a mentor but the research he started me on turned out to be most rewarding. More than any of his other students, I followed in his footsteps, continuing his work on atomic structure calculations using numerical methods. Computers were developing rapidly and there was great interest in quantum theoretical calculations, both in physics and chemistry. Then in 1960, I gained access to my first FORTRAN compiler. I saw this as a great opportunity in that the compiler facilitated portability from one computer to the next. I soon started on the development of what is now called the MCHF program (for Multi-Configuration Hartree Fock) based on the "superposition of configuration" work of Hartree, Hartree and Swirles. During the course of this work I found that Hartree's numerical methods worked better for the ground state than for excited states and modifications were needed. His elaborate procedures for getting the best possible initial estimates were unnecessary. With good computer codes, it was more convenient to use a simple but general starting procedure and let the computer iterate a bit longer. The importance of configuration interaction, was found to be critical for many properties, particularly for the prediction of electron affinities (the binding energy of an electron to a neutral atom). With configuration interaction, I was able to show the existence of a negative calcium ion. Whereas the first calculation by Hartree, Hartree, and Swirles, was a superposition of only two configuration states, today 10's of thousands may be included.

Now computers are so powerful that many tasks that I struggled over in the past are trivial. My Hartree Fock program can perform calculations for the ground state of radon with 86 electrons in a few seconds. Hartree prepared me for this astounding computing revolution. It indeed was a privilege to have been his student.

Notes and References

[1] In describing Sir Aaron Klug's work with Hartree, his Nobel Lecture, available on the web, was supplemented by letter, written November 8, 2000.

Hartree's publications

Books

[1] 1925 Contribution to *Textbook of anti-aircraft gunnery* Vol. 1 (H M Stationery Office)
[2] 1927 Substantial revision of English translation of M Born's *Vorlesungen über Atommechanik*. Translation entitled *The mechanics of the atom* (London: Bell & Co.)
[3] 1949 *Calculating Instruments and Machines* (Urbana: University of Illinois Press): 1950 English Edition (Cambridge: Cambridge University Press)
[4] 1952 *Numerical Analysis* (Oxford University Press): 1958 (2nd edition)
[5] 1957 *The calculation of atomic structures* (New York: Wiley & Sons)
[6] 1984 *Calculating Machines: Recent and prospective developments and their impact on Mathematical Physics and Calculating Instruments and Machines* (with a new introduction by Maurice V Wilkes) The Charles Babbage Institute reprint series for the history of computing, Vol. 6 (Cambridge: The MIT Press)

Papers

[1] Hartree D R 1912 Observations on certain periodic properties of numbers *J. Bedales School Sci. Soc.* No 2 29
[2] Bairstow L, Fowler R H and Hartree D R 1920 The pressure distribution on the head of a shell moving at high velocities *Proc. Roy. Soc. A* **97** 202-218
[3] Hartree D R 1920 Ballistic calculations *Nature* **106** 152-154
[4] Hartree D R 1923 On the propagation of certain types of electromagnetic waves *Phil. Mag. Ser. 6* **46** 454-460
[5] Hartree D R 1923 On some approximate numerical applications of Bohr's theory of spectra *Proc. Camb. Phil. Soc.* **21** 625-641
[6] Hartree D R 1923 On the correction for non-uniformity of field in experiments on the magnetic deflection of β-rays *Proc. Camb. Phil. Soc.* **21** 746-752
[7] Hartree D R 1923 On atomic structure and the reflexion of X-rays by crystals *Phil. Mag. Ser. 6* **46** 1091-1111
[8] Hartree D R 1923-25 The spectra of some lithium-like and sodium-like

atoms *Proc. Camb. Phil. Soc.* **22** 409-425

[9] Hartree D R 1923-25 Some methods of estimating the successive ionization potentials of any element. *Proc. Camb. Phil. Soc.* **22** 464-474

[10] Hartree D R 1924 Some relations between optical spectra of different atoms of the same electronic structure. I – Lithium-like and Sodium-like atoms *Proc. Roy. Soc. A* **106** 552-580

[11] Hartree D R 1925 Note on Dr Turner's paper "Quantum defect and atomic number" *Phil. Mag. Ser. 6* **49** 390-396

[12] Hartree D R 1923-25 Doublet and triplet separations in optical spectra as evidence whether orbits penetrate into the core *Proc. Camb. Phil. Soc.* **22** 904-918

[13] Hartree D R 1923-25 The atomic structure factor in the intensity of reflexion of X-rays by crystals 1923-25 *Phil. Mag. Ser. 6* **50** 289-306

[14] Hartree D R 1925 The ionization potential of ionized manganese *Nature* **116** 356

[15] Fowler R H and Hartree D R 1926 An interpretation of the spectrum of ionized oxygen (O II) *Proc. Roy. Soc. A* **111** 83-94

[16] Hartree D R 1926 Some relations between the optical spectra of different atoms of the same electronic structure. II – Aluminum-like and Copper-like atoms *Proc. Camb. Phil. Soc.* **23** 304-326

[17] Hartree D R 1928 The wave mechanics of an atom with a non-coulomb central field. Part I – Theory and Methods *Proc. Camb. Phil. Soc.* **24** 89-110

[18] Hartree D R 1928 The wave mechanics of an atom with a non-coulomb central field. Part II – Results and discussion *Proc. Camb. Phil. Soc.* **24** 111- 132

[19] James R W, Waller I and Hartree D R 1928 An investigation into the existence of zero-point energy in the rocksalt lattice by an X-ray diffraction method *Proc. Roy. Soc. A* **118** 334-350

[20] Hartree D R 1928 The wave mechanics of an atom with a non-Coulomb central field. Part III – Term values and intensities in series in optical spectra *Proc. Camb. Phil. Soc.* **24** 426-437

[21] Hartree D R 1928-29 The propagation of electromagnetic waves in a stratified medium *Proc. Camb. Phil. Soc.* **25** 97-120

[22] Hartree D R 1929 The distribution of charge and current in an atom consisting of many electrons obeying Dirac's equations *Proc. Camb. Phil. Soc.* **25** 225-236

[23] Hartree D R 1929 The wave mechanics of an atom with a non-coulomb central field. Part IV – Further results relating to terms of the optical spectrum *Proc. Camb. Phil. Soc.* **25** 310-314

[24] Waller I and Hartree D R 1929 On the intensity of total scattering by X-rays *Proc. Roy. Soc. A* **124** 119-142

[25] Hartree D R 1929 Die Elektrizitätsverteilung im Atom *Physik Zeitschrift* **30** 517-518

[26] Hartree D R 1930-31 The propagation of electromagnetic waves in a refracting medium in a magnetic field *Proc. Camb. Phil. Soc.* **27** 143-162

[27] Hartree D R 1931 Optical and equivalent paths in a stratified medium, treated from a wave standpoint *Proc. Roy. Soc. A* **131** 428-450
[28] Hartree D R and Miss M M Black 1933 A theoretical investigation of the oxygen atom in various states of ionization *Proc. Roy. Soc. A* **139** 311-335
[29] Hartree D R and Ingman A L 1932-33 An approximate wave function for the normal helium atom *Mem. Proc. Manchester Lit. Phil. Soc.* **77** 69-90
[30] Hartree D R 1932-33 A practical method for the numerical solution of differential equations *Mem. Proc. Manchester Lit. Phil. Soc.* **77** 91-107
[31] Hartree D R 1933 The dispersion formula for an ionized medium *Nature* **132** 929-930
[32] Hartree D R 1933 Results of calculations of atomic wave functions. I – Survey and self-consistent fields for Cl^- and Cu^+ *Proc. Roy. Soc. A* **141** 282-301
[33] Hartree D R 1934 Results of calculations of atomic wave functions. II – Results for K^+ and Cs^+ *Proc. Roy. Soc. A* **143** 506-517
[34] Hartree D R 1934 Approximate wave functions and atomic field for mercury *Phys. Rev.* **46** 738-743
[35] Hartree D R, de L Kronig R and Petersen H 1934 A theoretical calculation of the fine structure for the K-absorption band of Ge in $GeCl_4$ *Physica* **1** 895-924
[36] Hartree D R 1935 The bearing of statistical and quantum mechanics on school work *Math. Gaz.* **19** 73-78
[37] Hartree D R and Hartree W 1935 Results of calculations of atomic wave functions. III – Results for Be, Ca, and Hg *Proc. Roy. Soc. A* **149** 210-231
[38] Hartree D R and Hartree W 1935 Self-consistent field, with exchange, for beryllium *Proc. Roy. Soc. A* **150** 9-33
[39] Hartree D R 1935 The differential analyzer *Nature* **135** 940-943
[40] Hartree D R and Porter A 1934-5 The construction and operation of a model differential analyzer *Mem. Proc. Manchester Lit. Phil. Soc.* **79** 51-72
[41] Hartree D R 1935 Results of calculations of atomic wave functions. IV – Results for F^-, Al^{+3}, and Rb^+ *Proc. Roy. Soc. A* **151** 96-105
[42] Hartree D R 1935-36 Some properties and applications of the repeated integrals of the error function *Mem. Proc. Manchester Lit. Phil. Soc.* **80** 85-102
[43] Callender A, Hartree D R and Porter A 1936 Time-Lag in a control system *Phil. Trans. Roy. Soc. A* **235** 415-444
[44] Nuttall A K, Hartree D R and Porter A 1936 The response of a non-linear electric circuit to an impulse *Proc. Camb. Phil. Soc.* **32** 304-320
[45] Hartree D R and Hartree W 1936 Self-consistent field, with exchange, for beryllium II - The (2s)(2p) ^3P and ^1P excited states *Proc. Roy. Soc. A* **154** 588-607
[46] Hartree D R and Hartree W 1936 Self-consistent field, with exchange, for Cl^- *Proc. Roy. Soc. A* **156** 45-62
[47] Hartree D R, Chapman S and Milne E A 1936 [Vector notation] *Math.*

Gaz. **20** 272-275
[48] Hartree D R 1936 The application of the differential analyzer to the solution of partial differential equations *Rep. Int. Congr. Mathematicians* Oslo
[49] Hartree D R and Hartree W 1936 Self-consistent field, with exchange, for Cu^+ *Proc. Roy. Soc. A* **157** 490-502
[50] Hartree D R 1936 The theory of complex atoms *Nature* **138** 1080-1082
[51] Myers D M, Hartree D R and Porter A 1937 The effect of space-charge on the secondary current in a triode *Proc. Roy. Soc. A* **158** 23-37
[52] Hartree D R 1937 On an equation occurring in Falkner and Skan's approximate treatment of the equation of the boundary layer *Proc. Camb. Phil. Soc.* **33** 223-239
[53] Hartree D R and Swirles Bertha 1937 The effect of configuration interaction on the low terms of the spectra of oxygen *Proc. Camb. Phil. Soc.* **33** 240-249
[54] Hartree D R 1936-37 Note on a set of solutions of the equation $y'' + (2/x)y' - y^2 = 0$ *Mem. Proc. Manchester Lit. Phil. Soc.* **81** 19-27
[55] Hartree D R and Womersley J R 1937 A method for the numerical or mechanical soution of certain types of partial differential equations *Proc. Roy. Soc. A* **161** 353-366
[56] Hartree D R, Porter A, Callender A and Stevenson A B 1937 Time-lag in a Control Systems II *Proc. Roy. Soc. A* **161** 460-476
[57] Hartree D R and Hartree W 1938 Self-consistent field, with exchange, for calcium *Proc. Roy. Soc. A* **164** 167-191
[58] Hartree D R and Hartree W 1938 Self-consistent field, with exchange, for potassium and argon *Proc. Roy. Soc. A* **166** 450-464
[59] Hartree D R and Hartree W 1938 Wave functions for negative ions of sodium and potassium *Proc. Camb. Phil. Soc.* **34** 550-558
[60] Hartree D R 1938 The mechanical integration of differential equations *Math. Gaz.* **22** 342-364
[61] Hartree D R and Nuttall A K 1938 The differential analyser and its application in electrical engineering *J. Inst. Electrical Engineers* **83** 643-647
[62] Hartree D R and Porter A 1938 The application of the differential analyzer to transients on a distortionless transmission line *J. Inst. Electical Engineers* **83** 648-656
[63] Hartree D R and Ingham J 1938-39 Note on the application of the differential analyzer to the calculation of train running times *Mem. Proc. Manchester Lit. Phil. Soc.* **83** 1-15
[64] Hartree D R, Hartree W and Swirles B 1939 Self-consistent field, including exchange and superposition of configurations, with some results for oxygen *Phil. Trans. A* **238** 229-247
[65] Copple C, Hartree D R, Porter A and Tyson H 1939 The evaluation of transient temperature distributions in a dielectric in an alternating field *J. Inst. Electrical Engineers* **85** 56-66
[66] Hartree D R 1938-39 Note on an integral occurring in the cascade theory of cosmic ray showers *Mem. Proc. Manchester Lit. Phil. Soc.* **83** 175-182

[67] Hartree D R and Johnston S 1938-39 Note on the function $\chi(x) = \int_0^\infty exp(-(x-w^2)^2)dw$ *Mem. Proc. Manchester Lit. Phil. Soc.* **83** 183-188
[68] Crank J, Hartree D R, Ingham J and Sloane R W 1939 Distribution of potential in cylindrical thermionic valves *Proc. Phys. Soc.* **81** 952-971
[69] Hartree D R 1939 A solution of the laminar boundary layer equations for retarded flow *Aero. Res. Cttee, R and M* No 2426
[70] Hartree D R 1939 The solution of the equations of the laminar boundary layer for Schubauer's observed pressure distribution for an elliptic cylinder *Aero. Res. Cttee, R and M* No 2427
[71] Hartree D R 1940 A great calculating machine: the Bush differential analyser and its applications in science and industry *Proc. Royal Institution* **31** 151-1 194
[72] Hartree D R 1940 The Bush differential analyser and its applications *Nature* **146** 319-323
[73] Hartree W, Harteee D R and Manning M F 1941 Self-consistent field calculations for Zn, Ga, Ga^+, Ga^{+++}, As, As^+, As^{++}, As^{+++} *Phys. Rev.* **59** 299-305
[74] Hartree W, Hartree D R and Manning M F 1941 Self-consistent field calculations for Ge^{++} and Ge *Phys. Rev.* **59** 306-307
[75] Hartree W, Hartree D R and Manning M F 1941 Self-consistent field, with exchange, for Si IV and Si V *Phys. Rev.* **60** 857-865
[76] Hartree D R 1943 Mechanical integration in electrical problems *J. Inst. Electrical Engineers* **90** 435-442
[77] Jackson R, Sarjant R J, Wagstaff J B, Eyres N R, Hartree D R and Inghan J 1944 Variable heat flow in steel *J. Iron and Steel Institute* **150** 211-267
[78] Eyres N R, Hartree D R, Ingham J, Jackson R, Sarjant R J and Wagstaff J B 1946 The calculation of variable heat flow in solids *Phil. Trans. Roy. Soc. A* **240** 1-57
[79] Hartree D R 1946 The ENIAC, an electronic calculating machine *Nature* **157** 527
[80] Hartree D R 1946 The ENIAC, an electronic calculating machine *Nature* **158** 500-506
[81] Hartree D R 1946 The application of the differential analyzer to the solution of differential equations *Proc. Canadian Math. Congress* Montreal, 327
[82] Hartree D R, Michel J G L and Nicolson P 1947 Practical methods for the solution of the equations of tropospheric refraction *Meteorological factors in radio-wave propagation: report of a conference held on April 1946 at the Royal Institute of London* (Physical Society and Royal Meteorological Society, Great Britain) p 127-168
[83] Hartree D R 1947 Calculating machines, recent and prospective developments (Inaugural Lecture) (Cambridge Univ. Press)
[84] Hartree D R 1947 Recent developments in calculating machines *J. Sci. Instrum.* **24** 172-176
[85] Hartree D R 1947 Recent developments in calculating machines *Handel v*

h XXXe Nederlandisch Naturen Geneeskundig Congres Delft
[86] Hartree D R 1947 Recent and prospective developments in large digital calculating machines *J. Roy. Naval Sci. Service* (July)
[87] Hartree D R and Johnston S 1948 On a function associated with the logarithmic derivative of the gamma function *Q. J. Mech. Appl. Math.* **1** 29-34
[88] Hartree D R and the late Hartree W 1948 Self-consistent field, with exchange, for nitrogen and sodium *Proc. Roy. Soc.* A **193** 299-304
[89] Cope W F and Hartree D R 1948 The laminar boundary layer in compressible flow *Phil. Trans. Roy. Soc.* A **241** 1-69
[90] Hartree D R 1948 Experimental Arithmetic *Eureka* **10** 13
[91] Hartree D R 1948 A historical survey of digital computing machines (Contribution to discussion on Computing Machines) *Proc. Roy. Soc.* A **195** 265-271
[92] Hartree D R 1948 The calculation of atomic structure *Rep. Prog. Phys.* **11** 113-143
[93] Hartree D R (Editor) 1949 The differential analyzer *Ministry of Supply, Permanent Records of Research and Development, Monograph* **17.502**
[94] Hartree D R 1949 Notes on iterative processes *Proc. Camb. Phil. Soc.* **45** 230-236
[95] Hartree D R 1949 Note on systematic rounding errors in numerical integration *J. Res. Nat. Bur. Stand.* **42** 62
[96] Hartree D R 1949 The tabulation of Bessel functions for large argument *Proc. Camb. Phil. Soc.* **45** 554-55
[97] Hartree D R 1949 Modern Calculating Machines *Endeavour* **8** No. 30
[98] Hartree D R 1950 Some calculations of transients in an electronic valve *Appl. Sci. Res.* B **1** 379
[99] Hartree D R 1950 A method for the numerical integration of first-order differential equations *Proc. Camb. Phil. Soc.* **46** 523-524
[100] Hartree D R 1950 Automatic calculating machines *Math. Gaz.* **34** 241-252
[101] Hartree D R 1951 Automatische Rechenmaschinen (translation of above) *Z. Angew. Math. Mechanik* **31** 1
[102] Hartree D R 1951 Some unsolved problems in numerical analysis *Problems for the Numerical Analysis of the Future* National Bureau of Standards, Applied Mathematics Series 15, June 29, pp 1-9
[103] Hartree D R 1952 Automatic calculating machines and their potential application in the office *Office Management Assoc. J.* (August)
[104] Hartree D R 1954 The evaluation of a diffraction integral *Proc. Camb. Phil. Soc.* **50** 567-574
[105] Hartree D R 1955 Approximate wave functions, with exchange for Mn^{+2} *Proc. Camb. Phil. Soc.* **51** 126-130
[106] Douglas A S, Hartree D R and Runciman W A 1955 Atomic wave functions for gold and thallium *Proc. Camb. Phil. Soc.* **51** 486-503
[107] Hartree D R 1955 The interpolation of atomic wave functions *Proc. Camb. Phil. Soc.* **51** 684-692
[108] Hartree D R 1956 Approximate wave functions for the first long period *J.*

Opt. Soc. Am. **46** 350-353
[109] Froese Charlotte and Hartree D R 1957 Wave functions for the normal states of Ne^{+3} and Ne^{+4} *Proc. Camb. Phil. Soc.* **53** 663-668
[110] Hartree D R 1958 Variation of atomic wave functions with atomic number *Rev. Mod. Phys.* **30** 63-68
[111] Hartree D R 1958 Representation of the exchange terms in Fock's equations by a quasi-potential *Phys. Rev.* **109** 840-841
[112] Hartree D R 1958 A method for the numerical integration of the linear diffusion equations *Proc. Camb. Phil. Soc.* **54** 207-213
[113] The late Hartree D R 1960 The radial charge densities for the Ti^{+2} argon core *Proc. Camb. Phil. Soc.* **56** 174-175

Index

ACE, 150, 155, 165
Appleton E V, 66, 74, 136, 147
Appleton-Hartree equation, 69
atomic units, 34
automatic following, 127

Badley J H, 9
ballistics, 12, 126
Beard M, 182
Bedales, 9, 18, 65, 78, 157
Bethe H, 76
Beyer Professorship, 73
Birchoff, Garrett, 176
Black M M, 55, 56, 74
Blackett P M S, 15, 81, 145, 162
Bohr, Niels, 23, 39, 42, 47, 75
Bohr theory, 24
bomb project, 133
Boot A H, 136, 138
boundary layer problem, 109, 110, 166
Bowen E G, 136, 181
Bragg, W Lawrence, 38, 75, 161
Brigands, 12
Buneman Oscar, 74, 136, 140
Buneman-Hartree criterion, 141
Bush, Vannevar, 85, 95, 119, 137, 145

Caldwell S, 145, 153
Callender A, 104
cavity magnetron, 138
Charlton, Elaine, see Hartree, Elaine

computer terminology, 173
Comrie L J, 111, 153
configuration interaction, 57
continuum functions, 44
control law, 105
Copley, David, 136
Copple C, 107
Crank J, 107, 135, 204
Crank-Nicolson formula, 135

Darwin, Charles G, 59, 68, 146
Dederick L S, 167
differential analyzer, 85, 123, 124, 139, 156, 173
  Manchester machine, 92
  Meccano model, 90
differential equations, 86, 97, 105, 110, 126
diffusion equation, 133
Dirac P A M, 33, 59, 161
Douglas A S, 188

Eckert J P, 148
EDSAC, 62, 153, 177, 185, 187
  atomic calculations
    relativistic, 188
    with exchange, 187, 189
    without exchange, 188
  computing on, 192
EDVAC draft report, 147, 153
effective nuclear charge, 25, 36, 69, 190

Ehrenfest, Paul, 28
Einstein, Albert, 23, 28
electronic brain, 154, 165
ENIAC
  benchmark calculation, 166, 167, 174
  computer, 146, 152
  exchange, 50, 52, 55, 56
  exclusion principle, 39, 48
Eyres, Nicholas R, 124, 134, 205

Falkner V M, 109
family tree, xv
Fock equations, 53, 56, 190
Fock V A, 50, 165
Fowler R H, 11, 24, 147, 161, 201
Frisch, Robert, 133
Froese C, 57, 208
Fuchs, Klaus, 134

Gamow G, 43
Garstang, Roy H, 169, 187, 207
Gaunt J A, 37
Gill S, 177, 185, 188
Gillon P N, 149, 167
Gold M T, 187
Goldberg L, 56
Goldstein, Sydney, 68, 109
Goldstine H H, 149, 153, 158, 167, 175
Goudsmit, Samuel, 28

Hargreaves J, 37, 52
Hartree
  Colin, 14
  Elaine, 18, 77, 81, 120, 124, 166
  Eva, 5, 6, 165
  Isabella Charlotte, 2, 5
  John Edwin, 14
  John Penn, 2
  Margaret, 18, 118, 119
  Oliver, 118
  Richard, 80, 118, 121, 180
  William, 1, 7, 15, 55, 118, 165
  William Sr, 2
Hartree condition, 141

Hartree equation, 141
Hartree harmonics, 140
Hartree House, 159
Harvard Mark I, 147
Hazen, Harold, 86, 103
heat flow, 130, 135, 206
high speed fading, 128
Hill A V, 7, 11, 18, 137
Howlett, Jack, 73, 123, 124, 126, 134, 204
Huskey, Harry D, 155
Hylleraas E, 41

inaugural address, 163
Ingham J, 107, 124, 127
initial estimates, 190
Innes F I, 57
instability voltage, 141
isotope separation, 133
iteration process, 175

James R W, 205
Jeffreys, Bertha, see Swirles, Bertha
Jeffreys, Harold, 201
Jucys A P, 58

Klug, Aaron, 205
Kyhl R L, 137

laminar flow, 109
Lehmer D H, 167
Leigh D C F, 179, 207
Lennard-Jones J E, 98, 153, 161
Lindsay, Robert Bruce, 29, 45, 62, 82
Lockett, Phyllis, 124, 129, 135, 136, 140
Lomax R J, 209
London, Fritz, 40, 76
Lovell B, 81
Lyons & Co., 158

machine's-eye view, 166
Madsen J P V, 184
magnetron
  cavity, 138, 139
  cylindrical, 138

magnetron research, 136
Mathematical Laboratory, 153
Mathematical Tables Committee, 197
mathematics curriculum, 179
Mauchly J W, 148
Mayers, David F, 62, 188, 208
McDougall J, 53
McDougall R, 93
McNulty K, 167
Meccano, 90
Metropolitan-Vickers Electrical Co., Ltd, 107
Michel J G L, 126, 129, 156
Milne E A, 11, 73
MIT Radiation Laboratory, 137, 142
Moore School, University of Pennsylvania, 162, 167
Mott N F, 42, 75, 162, 172
Mountbatten's speech, 153
music, 50, 80, 169
Myers, David M, 95, 181

NBS Inst of Numerical Analysis, 171
negative ions, 56
Nicolson M M, 127
Nicolson, Phyllis, see Lockett, Phyllis
NPL Executive Committee, 147, 150, 155
numerical analysis, 178
   unsolved problems, 171

obituaries, 198
Oliphant M L E, 136, 184
Ovenstone J A, 179

partial differential equations, 129, 179
particle simulation, 139
Pearcey T, 182
Peierls R E, 76, 78, 133, 162
Penn
   Charlotte, 2
   John, 1
Petrashen M, 165
photo-electric effect, 69
Plummer Professorship, 152, 161

Porter, Arthur, 74, 95, 124, 128, 145, 163, 201

radio propagation, 128
radio waves, 65
Randall J T, 136, 138
Rayner, Eva, see Hartree, Eva
relativistic theory, 59
relativistic wave functions, 62
Ridley, Cicely, 188
Rosseland S, 92
Rutherford, Ernest, 23, 59

Sayers J, 136
Schrödinger's equation, 34
self-consistency, 30, 45
self-consistent field, 36, 47, 86, 91, 139
Self-help, 3
separation point, 111
Servo Panel, 123
Simmons J R M, 158
Skan S W, 109
Slater integrals, 54
Slater, John C, 47, 52, 85, 115, 137, 141, 148, 152, 172, 195
Sloane R W, 107
Smiles
   Janet, 4
   Samuel, 3, 166
Smith Newbern, 68
space-charge, 181
spin-orbitals, 48
superposition of configurations, 57
SWAC, 156
Swirles, Bertha, 19, 57, 61, 76, 118, 187, 201, 207
Sydney conference, 183
symmetry properties, 60

time-lag, 92, 104
train running times, 97
Turing A, 148, 150, 154, 155
Tyson H, 107

variational principle, 49, 50
von Neumann J, 153, 175

Waller Ivar, 38, 67
Watson W H, 195
wave propagation, 65, 66, 128
Wheeler D J, 177, 185
White F W G, 182
Whyte L L, 11, 28
Wilkes M V, 62, 98, 153, 158, 162, 177, 185
Womersley J R, 110, 147, 155
Worsley, Beatrice H, 187, 193

X-ray reflection, 26, 38